Rings,
Modules and
Linear Algebra

Rings, Modules and Linear Algebra

A further course in algebra describing the structure of Abelian groups and canonical forms of matrices through the study of rings and modules

B. HARTLEY
*Lecturer in Mathematics
University of Warwick*

T. O. HAWKES
*Lecturer in Mathematics
University of Warwick*

LONDON
CHAPMAN AND HALL

First published 1970
Reprinted 1974 and 1976
© B. Hartley, T. O. Hawkes, 1970
Printed in Great Britain at the
University Printing House, Cambridge

ISBN 0 412 09810 5

This limp-bound edition is sold subject to the condition that it shall not, by way of trade or otherwise, be lent, re-sold, hired out, or otherwise circulated without the publisher's prior consent in any form of binding or cover other than that in which it is published and without a similar condition including this condition being imposed on the subsequent purchaser.

All rights reserved. No part of this book may be reprinted, or reproduced or utilized in any form or by any electronic, mechanical or other means, now known or hereafter invented, including photocopying and recording, or in any information storage and retrieval system, without permission in writing from the Publisher.

Distributed, in the U.S.A. by
Halsted Press, a Division
of John Wiley & Sons, Inc.
New York

Preface

This book is based on a course of lectures given to mathematics undergraduates at the University of Warwick at the beginning of their second year. At this stage a student has completed a course in foundations, introducing the modern notation and basic structures now familiar to many school-leavers, and an elementary course in linear algebra. We have therefore assumed our reader has a well-developed facility with the language of sets, operations and mappings, as well as a working knowledge of vector spaces, linear transformations and matrices.

The object of this short text is to bring to a wide undergraduate audience a readable, leisurely and, at the same time, rigorous account of how a certain fundamental algebraic concept can be introduced, developed, and applied to solve some concrete algebraic problems. Among these problems are:

(a) How to classify finitely-generated Abelian groups.
(b) Given a linear transformation of a finite-dimensional vector space into itself, how to choose a basis so that the matrix associated with it has a particularly easy and manageable form.

The fundamental concept is that of a module over a ring; it is an idea of central importance in modern algebra and one which brings under the same roof many familiar ideas which may at first sight seem to be unconnected. By clamping down heavily on the type of ring allowed and imposing a further mild restriction we obtain modules for which a complete structure theory can be developed. For the applications we feed into this general theory some special cases which then become amenable to more detailed analysis.

The book is divided into three parts. The first is concerned with defining concepts and terminology, assembling elementary facts, and developing the theory of factorization in a principal ideal domain which we shall later need. The second part deals with the main decomposition theorems which describe the structure of finitely-generated modules over a principal ideal domain. The third, and in some ways the most important, part contains the applications of these theorems. One of these applications is the problem of classifying, up to a change of basis, the linear transformations of a vector space into itself. It turns out to be equivalent to that of finding canonical forms for matrices under similarity – in particular the Jordan form. This, of course, is a problem of considerable significance in its own right, and one which crops up time and again across the board of modern mathematics, from differential equations to projective geometry. The language of module theory provides an elegant and conceptually simple view of the Jordan form. At the same time it is itself a language of ever increasing importance in mathematics and one which should be introduced at an early stage – especially since in its elementary form it is essentially the theory of vector spaces over a general ring instead of a field, and as such finds its natural place in a 'second course in linear algebra'. Thus Parts II and III play a complementary rôle, the general theory of Part II exhibiting the conceptual unity and simplicity of the applications in Part III, while at the same time the applications provide a striking justification for the general theory and a sound concrete background to it. For further information about the organization of the book the reader should consult the flow chart on p. vii.

Organization of topics

The continuous path denotes the main route through the book. The dotted path is an alternative which does not extend to the bottom two topics.

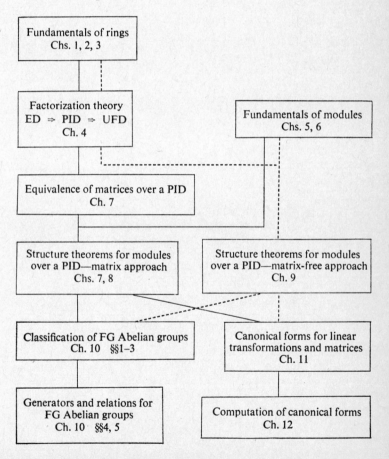

Notes for the reader:

1. Definitions, Lemmas, Theorems etc. are labelled consecutively with numbers of the form m.n where m denotes the chapter number and n the position within the chapter.
2. Where equations need to be referred back to, these are enumerated with a number (n) on the right-hand side of the page. The numbering begins again with each chapter.
3. Exercises are provided at the end of each chapter. Those marked with a star are intended to be more difficult.

Contents

Preface v

PART I RINGS AND MODULES

1. Rings – definitions and examples
  1. The definition of a ring 3
  2. Some examples of rings 5
  3. Some special classes of rings 11

2. Subrings, homomorphisms and ideals
  1. Subrings 15
  2. Homomorphisms 18
  3. Some properties of subrings and ideals . . . 26

3. Construction of new rings
  1. Direct sums 33
  2. Polynomial rings 37
  3. Matrix rings 44

4. Factorization in integral domains
  1. Integral domains 49
  2. Divisors, units and associates 51
  3. Unique factorization domains 54
  4. Principal ideal domains and Euclidean domains . 59
  5. More about Euclidean domains 62

5. Modules
  1. The definition of a module over a ring . . . 69
  2. Submodules 74
  3. Homomorphisms and quotient modules . . . 77
  4. Direct sums of modules 80

6. **Some special classes of modules**
   1. More on finitely-generated modules . . . 85
   2. Torsion modules 87
   3. Free modules 89

PART II DIRECT DECOMPOSITION OF A FINITELY-GENERATED MODULE OVER A PRINCIPAL IDEAL DOMAIN

7. **Submodules of free modules**
   1. The programme 99
   2. Free modules – bases, endomorphisms and matrices . 100
   3. A matrix formulation of Theorem 7·1 . . . 106
   4. Elementary row and column operations . . . 110
   5. Proof of 7.10 for Euclidean domains . . . 111
   6. The general case 114
   7. Invariant factors 115
   8. Summary and a worked example 118

8. **Decomposition theorems**
   1. The main theorem 123
   2. Uniqueness of the decomposition 127
   3. The primary decomposition of a module . . . 132

9. **Decomposition theorems – a matrix-free approach**
   1. Existence of the decompositions 139
   2. Uniqueness – a cancellation property of FG modules 143

PART III APPLICATIONS TO GROUPS AND MATRICES

10. **Finitely-generated Abelian groups**
    1. Z-modules 151
    2. Classification of finitely-generated Abelian groups . 152
    3. Finite Abelian groups 154
    4. Generators and relations 157
    5. Computing invariants from presentations . . 160

CONTENTS xi

11. Linear transformations, matrices and canonical forms
    1. Matrices and linear transformations . . . 167
    2. Invariant subspaces 169
    3. V as a $k[x]$-module 170
    4. Matrices for cyclic linear transformations . . 176
    5. Canonical forms 179
    6. Minimal and characteristic polynomials . . . 184

12. Computation of canonical forms
    1. The module formulation 193
    2. The kernel of ϵ 195
    3. The rational canonical form 197
    4. The primary rational and Jordan canonical forms . 200

References 205

Index 207

PART ONE

Rings and Modules

CHAPTER ONE

Rings – definitions and examples

1. The definition of a ring

A ring is a natural object of study because it crops up in so many varied and important mathematical contexts – this will become apparent from the examples which we shall list. The axioms defining a ring are derived from some of the important properties of the set **Z** of integers. In fact, the integers may be taken as a prototype for a ring; they will frequently appear as a source of motivation and examples. Like the integers a ring R is a set with two binary operations; these are usually called *addition* (denoted by the sign +) and *multiplication* (denoted by juxtaposition). R is then a ring if it forms a commutative group with respect to addition, a semigroup with respect to multiplication and satisfies distributive laws connecting the two operations.

We will be more precise. First we recall that a binary operation on a set S is a map $\mu: S \times S \to S$ where $S \times S$ denotes the Cartesian product of S with itself, that is the set of all ordered pairs (a,b) with a and b in S. We usually write the image under μ of the pair (a,b) as $a*b$, where $*$ is the appropriate symbol for the binary operation – for example, as we have said, the symbol + is used when the operation is called addition and is left out for multiplication. Note that the order is usually important in the sense that $a*b$ and $b*a$ may be different elements of S. However, if $a*b = b*a$ for all pairs of elements a and b in S, the operation $*$ is called *commutative*. In a similar spirit a map from S to itself is often called a *unary operation*.

1.1. Definitions. (a) A *semigroup* is a set S with a binary operation satisfying the associative law, that is

$$a * (b * c) = (a * b) * c \quad \text{for all } a, b, c \in S.$$

(b) A *group* is a *non-empty* set G with a binary operation $*$, a unary operation $x \to \bar{x}$, and a selected element $e \in G$ such that

(i) G is a semigroup with respect to $*$,
(ii) $a * e = e * a = a$ for all $a \in G$, and
(iii) $a * \bar{a} = \bar{a} * a = e$ for all $a \in G$.

The element e is called the *neutral element* or *identity element* of G, and $\bar{a}$ is called the *inverse* of a. It is almost invariable practice to use either multiplicative or additive notation for groups. In the former one leaves out the symbol $*$, writes a^{-1} for $\bar{a}$, and usually writes 1 instead of e. In additive notation the symbol $*$ is replaced by $+$, $\bar{a}$ is replaced by $-a$, and e is replaced by 0. Additive notation is usually (though not invariably) reserved for groups in which the binary operation is commutative. Such groups are usually called *Abelian* (rather than commutative) in honour of the distinguished Norwegian mathematician N. H. Abel (1802–1829) who investigated a class of algebraic equations related to commutative groups. In passing, we recall from elementary group theory the fact that the neutral element and the unary operation of inversion are both uniquely determined.

(c) A *ring* is a set R equipped with two binary operations which are connected by distributive laws. R is an Abelian group with respect to one binary operation (by convention called addition and denoted by $+$) and a semigroup with respect to the other binary operation (called multiplication and denoted by juxtaposition). Observe that R is therefore non-empty. The left and right distributive laws connecting these two operations are as follows:

$$\left. \begin{array}{l} a(b+c) = ab + ac \\ (a+b)c = ac + bc \end{array} \right\} \text{ for all } a, b, c \in R.$$

The reader may find it instructive at this point to write out all the axioms of a ring in detail.

It is clear that the integers (henceforth denoted by **Z**) with the usual definitions of addition and multiplication, do indeed satisfy

RINGS – DEFINITIONS AND EXAMPLES

the axioms of a ring just revealed. Fortunately these axioms by no means characterize **Z**, for if they did the 'theory of rings' would be a mathematical dead duck. This is not to belittle the study of the integers but merely to emphasize that a ring is a wide-ranging concept having many manifestations and embracing many different situations. To indicate how special **Z** is among the fraternity of rings we observe that **Z** has commutative multiplication, it has a multiplicative identity, it has an ordering, it is countable, it possesses a well-behaved factorization, and that none of these requirements has been included in our definition of a ring. This definition gives rise to a great variety of structures, as the list in the next paragraph shows.

2. Some examples of rings

To understand a general mathematical theory it is important to try it out on some concrete, and if possible familiar, examples. Often the real significance of a theorem only becomes clear after its implications have been explored in some simple test cases. Hence the value of having readily accessible a varied repertoire of examples of the structure under consideration – in this case a ring. What other aids to understanding a proof are there? The statement of a theorem usually consists of a collection of hypotheses followed by some consequences or conclusions, and an undoubtedly virtuous activity for a student wading through the details of a proof is to pinpoint exactly where each of the hypotheses is used. He is then led to ask whether the theorem remains true under weaker hypotheses, and the resulting investigation may involve a search for counter-examples to prove that under relaxed assumptions the conclusions of the theorem are no longer true. Here again a well-filed mental list of examples is very helpful. Hence we emphasize the rôle of examples in this book. We begin here by giving a short list of examples of rings which will frequently be referred back to; in the next two chapters we will give some general methods of constructing new rings from given ones.

Ring Example 1. If $n \in \mathbf{Z}$, the subset

$$n\mathbf{Z} = \{a \in \mathbf{Z} : n \text{ divides } a\}$$

of the integers is closed under addition and multiplication. It obviously satisfies the ring axioms, and therefore forms a ring in its own right.

Ring Example 2. Let n be a fixed positive integer, and define an equivalence relation $\sim$ on $\mathbf{Z}$ by: $a \sim b$ if and only if $a - b$ is divisible by n. Denote the equivalence class containing a by $[a]$. Then it is not difficult to see that $[0], [1], \ldots, [n-1]$ is a complete set of equivalence classes with respect to $\sim$. In other words, no two of the set $\{0, 1, \ldots, n-1\}$ are equivalent and every integer is equivalent to some member of that set. The equivalence classes just described are usually called the *congruence classes modulo n* or the *residue classes modulo n*, and the set of them is denoted by $\mathbf{Z}_n$. It turns out that if we define addition and multiplication of classes in terms of representatives (i.e. $[a] + [b] = [a+b]$ and $[a][b] = [ab]$), then these operations are well-defined and convert the set $\mathbf{Z}_n$ into a ring. This ring has finitely many, namely n, elements. We shall give a more general proof of these statements in Chapter 2 in the context of quotient rings. Meanwhile the reader might like to familiarize himself with one of these rings by writing out in detail the addition and multiplication tables of $\mathbf{Z}_6$ and convincing himself that the ring axioms hold. For example, in $\mathbf{Z}_6$ we have $[3] + [5] (= [8]) = [2]$, and $[3][5] (= [15]) = [3]$.

Ring Example 3. We can make any Abelian group A into a ring by writing it additively and defining multiplication by $ab = 0$ for all $a, b \in A$. We leave it as an exercise to check that with these definitions A satisfies the ring axioms.

Ring Example 4. The set $\mathbf{C}$ of complex numbers forms a ring under the usual operations of addition and multiplication. In fact, it is a ring and much more, for multiplication is commutative, there is a multiplicative identity, division by non-zero elements is possible, etc. It is easy to verify that the subsets $\mathbf{R}$ and $\mathbf{Q}$ of $\mathbf{C}$, denoting respectively the real and the rational numbers, form rings under the usual operations.

Ring Example 5. Another subset of $\mathbf{C}$, namely

$$J = \{a + ib : a, b \in \mathbf{Z}\},$$

RINGS – DEFINITIONS AND EXAMPLES

also forms a ring under addition and multiplication of complex numbers. It is easy to see that the sum, product and difference of two elements of J belongs again to J, and it follows almost immediately that J satisfies the ring axioms, once we know that these hold in $\mathbf{C}$. J is usually called the *ring of Gaussian integers*.

Ring Example 6. Given a set X, let $\mathscr{P}(X)$ denote the set of all subsets of X (including X itself and the empty set $\varnothing$). $\mathscr{P}(X)$ is usually called the *power set* of X. If X is finite and has n elements, then $\mathscr{P}(X)$ has 2^n elements. This is because in constructing a subset of X each element of X gives rise to two possibilities, according as the element is put into the subset or left out of it; thus the total number on subsets is 2^n. Rather surprisingly we can always put a ring structure of a power set in the following way. For any $A, B \in \mathscr{P}(X)$, that is for any subsets A and B of X, define

$$A + B = (A \cup B) \setminus (A \cap B) \quad \text{(the 'disjoint union')},$$

and

$$AB = A \cap B.$$

Here $C \setminus D$ denotes the set of elements which belong to C but not to D. These definitions satisfy the ring axioms. For example, $A + \varnothing = (A \cup \varnothing) \setminus (A \cap \varnothing) = A \setminus \varnothing = A = \varnothing + A$, so that $\varnothing$ is the additive identity, i.e. the zero. Also, $A + A = (A \cup A) \setminus (A \cap A) = A \setminus A = \varnothing$. So A is its own additive inverse, that is, $A = -A$. We leave the rest of the checking as an exercise. Some of the axioms are obvious, some need a few moments' thought, but all become clearer when visualized is terms of the appropriate Venn diagrams:

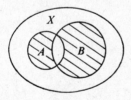
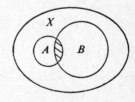

$A + B$ $\qquad\qquad\qquad\qquad AB$

Notice that when multiplication is commutative (as in this case) either of the distributive laws implies the other, and so only one of them need be checked.

Ring Example 7. Let $\mathbf{M}_n(\mathbf{k})$ denote the set of $n \times n$ matrices with entries in some field $\mathbf{k}$. The reader may think of $\mathbf{k}$ as the real numbers if he so desires. We recall the usual definitions of addition and multiplication on $\mathbf{M}_n(\mathbf{k})$. If $A = (a_{ij})$ and $B = (b_{ij})$ are elements of $\mathbf{M}_n(\mathbf{k})$, the (i, j) entry of $A + B$ is $a_{ij} + b_{ij}$ and of AB is $\sum_{k=1}^{n} a_{ik} b_{kj}$. With these binary operations $\mathbf{M}_n(\mathbf{k})$ forms a ring; it is intimately related to another ring which the reader has probably met, namely the ring of linear transformations of an n-dimensional vector space over $\mathbf{k}$. We shall recall this relationship in detail later. If $n > 1$, the ring $\mathbf{M}_n(\mathbf{k})$ has a non-commutative multiplication, unlike Examples 1–6. The reader can see this by considering the two matrices

$$\begin{bmatrix} 0 & 1 \\ 0 & 0 \end{bmatrix}, \quad \begin{bmatrix} 1 & 0 \\ 0 & 0 \end{bmatrix}$$

and obvious analogues of them.

Ring Example 8. For any set X (even the empty set, but exclude it if it troubles you) the set of all maps $f: X \to \mathbf{R}$ becomes a ring if we define

addition by: $(f + g)(x) = f(x) + g(x)$

multiplication by: $(fg)(x) = f(x) g(x)$.

This is sometimes called the pointwise definition of addition and multiplication; it uses the ring structure of $\mathbf{R}$ to induce a ring structure on the set of maps. We leave the reader to fill in the details (and to generalize?). If X itself is the set $\mathbf{R}$, then other rings can be obtained in this way; for example the set of continuous maps $\mathbf{R} \to \mathbf{R}$, the set of differentiable maps $\mathbf{R} \to \mathbf{R}$, etc., all form rings under the pointwise operations just described.

Ring Example 9. Let **1, i, j, k** denote the four elements

$$\mathbf{1} = \begin{bmatrix} 1 & 0 \\ 0 & 1 \end{bmatrix}, \quad \mathbf{i} = \begin{bmatrix} i & 0 \\ 0 & -i \end{bmatrix}, \quad \mathbf{j} = \begin{bmatrix} 0 & 1 \\ -1 & 0 \end{bmatrix}, \quad \mathbf{k} = \begin{bmatrix} 0 & i \\ i & 0 \end{bmatrix}$$

of $\mathbf{M}_2(\mathbf{C})$. Notice that here we are 'abusing notation' in several places by using the same symbol to denote two different things. For example, 1 denotes the complex number 1 as well as the 2×2 identity matrix over $\mathbf{C}$, although different printers' type is

RINGS – DEFINITIONS AND EXAMPLES

used to distinguish them. This improper practice is frequently necessary in mathematics if one wants to avoid being submerged in a welter of notation, but it is essential to realize when it is going on.

Let V denote the set of all elements of $\mathbf{M}_2(\mathbf{C})$ of the form

$$\mathbf{x} = a\mathbf{1} + b\mathbf{i} + c\mathbf{j} + d\mathbf{k} \quad (a, b, c, d \in \mathbf{R}). \tag{1}$$

Thus a general element of V has the form

$$\begin{bmatrix} a+bi & c+di \\ -c+di & a-bi \end{bmatrix} \quad (a, b, c, d \in \mathbf{R}).$$

It is straightforward to verify that the matrices $\mathbf{1}, \mathbf{i}, \mathbf{j}, \mathbf{k}$ multiply together according to the rules

$$\left. \begin{array}{l} \mathbf{i}^2 = \mathbf{j}^2 = \mathbf{k}^2 = -\mathbf{1}; \quad \mathbf{ij} = -\mathbf{ji} = \mathbf{k} \quad \text{and two similar equations} \\ \text{obtained by permuting } \mathbf{i}, \\ \mathbf{j} \text{ and } \mathbf{k} \text{ cyclically.} \end{array} \right\} \tag{2}$$

Hence it follows from the rules for matrix manipulation that the sum and product of any two elements of V belongs to V, and that if $\mathbf{x} \in V$ then $-\mathbf{x} \in V$. Therefore the ring operations in $\mathbf{M}_2(\mathbf{C})$ determine corresponding operations on V. The ring axioms will clearly be satisfied, and so these operations make V into a ring; V is a *subring* of $\mathbf{M}_2(\mathbf{C})$ in a sense to be made precise later. V is called the *ring of quaternions*.

If $\mathbf{x}$ is as in (1), let

$$\bar{\mathbf{x}} = a\mathbf{1} - b\mathbf{i} - c\mathbf{j} - d\mathbf{k}.$$

$\bar{\mathbf{x}}$ is called the *conjugate* quaternion to $\mathbf{x}$. The reader may verify by working out $\mathbf{x}\bar{\mathbf{x}}$ from the relations (2) that any non-zero matrix in V is actually non-singular and that its inverse belongs to V. In fact, if $\mathbf{x} \neq 0$, we have

$$\mathbf{x}^{-1} = \lambda \bar{\mathbf{x}},$$

where λ is the real number $1/(a^2 + b^2 + c^2 + d^2)$. Thus in V division by non-zero elements is always possible. However, multiplication in V is non-commutative, as the relations (2) show. So in some vague way the quaternions are just one step worse than the complex numbers. V contains several subsets which look very much like **C**, for example $\{a\mathbf{1} + b\mathbf{i}\}$, $\{a\mathbf{1} + b\mathbf{j}\}$, etc. The concept of isomorphism will later allow us to be more precise.

Ring Example 10. Let A be an arbitrary Abelian group, written additively. An *endomorphism* of A is a homomorphism of A into itself, in other words a map $\alpha: A \to A$ satisfying the condition $\alpha(a+b) = \alpha(a) + \alpha(b)$. The set End A of all endomorphisms of A can be given a ring structure in a natural way by defining

addition by: $(\alpha + \beta)(a) = \alpha(a) + \beta(a)$

multiplication by: $(\alpha\beta)(a) = \alpha(\beta(a))$

for all $a \in A$ and all $\alpha, \beta \in \text{End}\, A$. Thus the definition of addition is pointwise and of multiplication is by composition of maps. The reader should convince himself that this makes End A into a ring. Notice that the first step is to check that sums and products of endomorphisms are in fact endomorphisms to make sure that the above definitions give binary operations on End A. It is the commutativity of A which ensures that this is the case; it is false for groups in general.

Some 'non-examples'. It is a useful exercise to see why some possible candidates for the title of ring do not in fact live up to the axioms. We leave the reader to say why the following sets (with the obvious binary operations) do not fulfil the requirements of 1.1(c):

(a) The set of positive integers.

(b) The set of all rational numbers expressible in the form m/n, where m and n are integers and n is not divisible by 4.

(c) The subset of $\mathbf{M}_2(\mathbf{C})$ comprising matrices with zeros on the diagonal.

(d) The power set $\mathscr{P}(X)$ of a non-empty set X with addition redefined by

$$A + B = A \cup B$$

and multiplication unchanged.

(e) The set of $m \times n$ matrices ($m > n$) with entries in **C**.

(f) The set of 3-dimensional vectors with 'cross product' as the binary operation of multiplication.

RINGS – DEFINITIONS AND EXAMPLES

3. Some special classes of rings

As we have seen from the list of examples, the rings that crop up in real life often satisfy additional axioms to those required by 1.1(c). For this reason it is useful to pick out certain important special types of rings and give names to them. However, before doing this, we will get out of the way some elementary consequences of the ring axioms which are very often needed.

1.2. Lemma. *Let R be a ring. Then*

(i) $r0 = 0r = 0$,

(ii) $(-r)s = r(-s) = -(rs)$, *and*

(iii) $(-r)(-s) = rs$

for all $r, s \in R$.

Proof. (i) Since 0 is the additive identity, we have $0 + 0 = 0$. Hence $r(0 + 0) = r0$, and by one of the distributive laws $r0 + r0 = r0$. Therefore $r0 + r0 = r0 + 0$. By the cancellation law (which holds in any group) we find that $r0 = 0$. Similarly $0r = 0$.

(ii) We have $r + (-r) = 0$. Hence using (i) and the distributive law we find $(r + (-r))s = 0s = 0$, or $rs + ((-r)s) = 0$. But also by definition of $-(rs)$ we have $rs + (-(rs)) = 0$. Therefore by the cancellation law of addition we obtain $(-r)s = -(rs)$. A similar argument gives $r(-s) = -(rs)$.

(iii) By applying (ii) repeatedly we find that $(-r)(-s) = -(r(-s)) = -(-(rs))$. Now for any $t \in R$, $-t$ is the unique solution of the equation $t + x = 0$. Therefore the equation $(-t) + t = 0$ tells us that $-(-t) = t$. Hence $-(-(rs)) = rs$ and the proof is complete.

1.3. The generalized associative law. It is important to realize that the binary operation of multiplication in a ring only allows us to multiply two elements at a time. If we want to multiply three elements a, b, c in that order, we have to specify how it is to be done by inserting brackets, e.g. $a(bc)$, which means first calculate the product bc and then premultiply the resulting element by a. In the case of three elements there are only two ways in which the multiplication may be carried out, corresponding to $(ab)c$ and $a(bc)$. The associative law tells us that these give the same answer and so we may as well leave out the brackets. Whenever we evaluate a product abc in practice we always implicitly put in brackets somewhere, but the answer doesn't depend on where the brackets are put, and so the symbol abc has a unique meaning. Now the associative law as it stands does not tell us about the product $a_1 a_2 \ldots a_n$ of more than three elements. Does the symbol $a_1 a_2 \ldots a_n$ have a unique meaning? That is, whenever we put in brackets and evaluate, do we always get the same answer? The answer is yes, and can be deduced from the ordinary associative law. Since the reader is no doubt familiar with this kind of thing from elementary group theory, we shall not give the details. In fact, it is fairly difficult to give a really satisfactory proof, the difficulty lying mainly in formulating the problem properly, but the reader can easily get the idea of what is going on by experimenting with various bracketings of products of four and five elements and seeing how these bracketings can be transformed into one another by successive applications of the associative law. A precise account can be found in [3], p. 18. Similar remarks apply of course to addition, or to any associative binary operation for that matter.

We are now ready to introduce a few special classes of rings.

Commutative rings are rings in which multiplication is commutative, that is, rings in which the equation $ab = ba$ holds for arbitrary elements of a, b of the ring.

Rings with a multiplicative identity, usually called **rings with 1.** As the terminology suggests, a ring is in this class if it contains an element, usually denoted by 1, such that $r1 = 1r = r$ for all r in the ring. Notice that the ring $\{0\}$ comprising a single element is a ring with a 1, the '1' in question being 0 of course. The usual argument shows that in any ring with a 1 the '1' is uniquely determined. For let R be a ring with some multiplicative identity singled out and denoted by 1, and suppose that e is any multiplicative identity in R. Then $e = e1 = 1$, since both 1 and e are multiplicative identities.

Integral domains. An integral domain is a commutative ring which has a $1 \neq 0$ and which has no zero divisors. It remains to explain what the term 'zero divisor' means. Now the equation $r0 = 0$ established in Lemma 1.2 means that any element of any ring

RINGS – DEFINITIONS AND EXAMPLES

'divides zero'. Nevertheless, we define a *zero divisor* of a commutative ring R to be an element $r \in R$ such that

(i) $r \neq 0$, and
(ii) $rs = 0$ for some $s \neq 0$ in R.

(In dealing with non-commutative rings one has to distinguish between right and left zero divisors but since we shall be mainly concerned with commutative rings in this book we shall avoid getting caught up in these considerations.) The following fact about integral domains is important:

1.4. Lemma. *Let R be an integral domain, a a non-zero element of R and $x, y \in R$. Then $ax = ay \Rightarrow x = y$.*

This is known as the *cancellation law of multiplication*.

Proof. If $ax = ay$, then the ring axioms give $a(x - y) = 0$. Therefore, since a is not a zero divisor, we must have $x - y = 0$. Hence $x = y$.

Fields. A field is a commutative ring in which the set of non-zero elements forms a group under multiplication. Thus, if **k** is a field, then **k** contains an element $1 \neq 0$ such that $1x = x$ for all $x \neq 0$ in **k**. Since certainly $1 \cdot 0 = 0$ (by Lemma 1.2), 1 is a multiplicative identity. Furthermore, to each $a \neq 0$ in **k** there corresponds an element a^{-1} such that $aa^{-1} = 1$. It is easy to see that a field has no divisors of zero. For if $0 \neq a \in \mathbf{k}$, and $ax = 0$, then $x = 1x = a^{-1}ax = a^{-1}0 = 0$. Thus the following relations hold between the four kinds of rings which we have introduced:

$$\text{Field} \Rightarrow \text{Integral domain} \begin{array}{c} \nearrow \text{Commutative ring} \\ \searrow \text{Ring with 1} \end{array}$$

To shed some light on these definitions the reader should undertake two tasks: first find out to which classes Ring Examples 1–10 belong; second, find examples to show that no two of the classes are equal.

Exercises for Chapter 1

1. Is the set of integers a semigroup under the binary operation of subtraction?

2. Which of the following are rings?

(i) The set of all continuous functions $f: \mathbf{R} \to \mathbf{R}$ with addition defined pointwise and multiplication defined by composition of maps.

(ii) The set of all rational numbers expressible in the form a/b, where a and b are integers and p does not divide b. Here p denotes a fixed prime, and the operations are the usual ones.

(iii) The set of integers with a new addition $\dotplus$ and a new multiplication $\times$ defined in terms of the usual operations by
$$n \dotplus m = n + m + 1$$
$$n \times m = n + m + nm$$

3. Show that in the ring $\mathscr{P}(X)$, as defined in Ring Example 6, $x^2 = x$ for all x. In which of the rings $\mathbf{Z}_n$ is this true?

4. Let α and β be endomorphisms of a not necessarily Abelian group G, and define $\alpha + \beta$ by
$$(\alpha + \beta)(x) = \alpha(x)\beta(x)$$
for $x \in G$. Under what conditions is $\alpha + \beta$ an endomorphism? Give an example to show that these conditions are not always fulfilled.

5. Let R be an integral domain such that $x^2 = x$ for all $x \in R$. Show that R has exactly two elements.

6. Let R be a ring with a 1. Show that either $1 \neq 0$ or $R = \{0\}$.

7. Show that every integral domain with only finitely many elements is a field. (*Hint*: let $0 \neq a \in R$ and consider the map $x \to ax$ of R into itself. Show that it is injective and hence surjective.)

8. Let S be a set, R a ring and f a bijection $S \to R$. Define operations on S in terms of those on R by
$$\left. \begin{array}{l} s + s' = f^{-1}(f(s) + f(s')) \\ ss' = f^{-1}(f(s)f(s')) \end{array} \right\} s, s' \in S.$$
Show that S is a ring with these operations. Find operations on the set of positive integers which make that set into a ring.

CHAPTER TWO

Subrings, homomorphisms and ideals

As usual when one is confronted by some new kind of mathematical structure one's first instinct is to look for substructures and 'morphisms', or structure-preserving maps, of the structure in question. This will be the aim of the present chapter.

1. Subrings

2.1. Definition. A *subring* of a ring R is a subset S of R which is a ring under the operations which it inherits from R.

What does the above definition mean? First of all, the operations on R must determine operations on S. In the case of addition, for example, it means that the map $R \times R \to R$ given by $(a,b) \to a+b$, when restricted to $S \times S$, must give a map $S \times S \to S$; or in more everyday terms, that whenever a and b belong to S, then $a+b$ must also belong to S. Similarly ab and $-a$ also belong to S. Hence $a - b = a + (-b) \in S$. Another consequence which is easily forgotten is that since S is to be a group under $+$, S must be non-empty. We have therefore proved half of

2.2. Lemma. *Let S be a subset of a ring R. Then S is a subring of R if and only if (i) S is non-empty, and (ii) whenever $a, b \in S$ then $a - b \in S$ and $ab \in S$.*

Proof. We have proved the necessity of the above conditions. We must now prove the sufficiency. Since S is non-empty, it contains some element a. Therefore by (ii) it also contains the element $a - a = 0$. Hence we have $-b = 0 - b \in S$, and therefore $a + b = a - (-b) \in S$ also. Thus the two binary and one unary operation on R induce corresponding operations on S. The commutative and associative laws of addition are consequences of those in R, since when we add elements of S we do so by thinking of them as elements of R. Hence S is an Abelian group under $+$ and $-$ with identity element 0. The associative law of multiplication and the two distributive laws follow in the same way from those of R. Hence S is a ring.

Examples. 1. With the usual operations, each of **Z, Q, R, C** is a subring of the next one.
2. The ring of quaternions as considered in Ring Example 9, is a subring of $\mathbf{M_2(C)}$.
3. The set of all $n \times n$ matrices over a field **k** which have zeros below the main diagonal is a subring of $\mathbf{M_n(k)}$.

We now need to introduce a fairly substantial amount of notation – some familiar, some perhaps unfamiliar, but all useful.

Notation. 1. In dealing with a ring R, it is often useful to think of R simply as a group under the additive structure which it possesses, ignoring the multiplicative structure. When we want to emphasise that this is being done, we shall write R^+ instead of R. R^+ is called the *additive group* of R; strictly speaking R denotes a system consisting of a set, two binary operations and a unary operation on that set, and a selected element of the set. R^+ denotes the same system with the binary operation of multiplication omitted. Subgroups of R^+ are often called *additive subgroups* of R. An additive subgroup of R is thus a subset S of R which contains 0 and satisfies the condition: if $a, b \in S$ then $a - b \in S$.
2. If A is any Abelian group written additively, $a \in A$ and n is an integer, then na is defined by:

$na = a + \cdots + a$ (with n terms a) if $n > 0$
$0a = 0$
$na = (-n)(-a) = -(a + \cdots + a)$ (with $|n|$ terms) if $n < 0$.

If $a, b \in A$ and n, m are integers, then we have

$$n(a + b) = na + nb,$$
$$(n + m)a = na + ma,$$
$$(nm)a = n(ma),$$
$$1a = a.$$

SUBRINGS, HOMOMORPHISMS AND IDEALS

The reader is no doubt familiar with these elementary facts, and their proofs can in any case be found in any elementary book on group theory. In particular, since we can take A to be the additive group R^+ of a ring R, the above definitions hold good in any ring R. Here a word of caution is necessary – it is important to distinguish the operation $(n, a) \to na$ from ring multiplication in R, since in general n will not even be an element of R.

However, it sometimes happens that $\mathbf{Z}$ is identified with a subring of R in such a manner that the integer 1 plays the part of a multiplicative identity in R. In this case, if $n > 0$ and we multiply n by an element $a \in R$ as ring elements, we get $na = (1 + \cdots + 1)a = a + \cdots + a$ by the distributive law. Thus in this case the symbol na acquires the same meaning whether it is interpreted as a product of ring elements or by means of the above definition. Similarly so does $0a$ and $(-n)a$. Thus there is no possibility of confusion.

If a is an element of a ring R and n is a positive integer, we also define $a^n = a \ldots a$ (n times). We then have $a^{n+m} = a^n a^m$; $a^{nm} = (a^n)^m$ if $n, m > 0$. If R happens to have a 1, we furthermore define $a^0 = 1$ for $0 \neq a \in R$, and the identities just mentioned continue to hold.

3. Let S, T be arbitrary non-empty subsets of a ring R. We define

$$S + T = \{s + t : s \in S, t \in T\}$$

$$ST = \left\{ \sum_{i=1}^{n} s_i t_i : s_i \in S, t_i \in T, n = 1, 2, \ldots \right\}.$$

We are mainly interested in the above definitions in the case when S and T are additive subgroups of R and we make the definition in this form to ensure that the sum and product of two additive subgroups of R is again an additive subgroup.

2.3. Lemma. *Let R be a ring and S, T, U be non-empty subsets of R. Then*:

(i) $(S + T) + U = S + (T + U)$ *and* $(ST)U = S(TU)$.

(ii) *If S and T are additive subgroups of R, then so are $S + T$ and ST.*

(iii) *If S and T are subrings of R and R is commutative, then ST is a subring of R.*

Proof. (i) It is clear that $(S + T) + U = S + (T + U)$. Now since ST consists of all finite sums of elements of the form st with

$s \in S$ and $t \in T$, it is closed under addition. Therefore so are $(ST)U$ and $S(TU)$. Now an arbitrary element z of $(ST)U$ is a finite sum of elements of the form xu with $x \in ST$ and $u \in U$. Therefore x is a finite sum of elements of the form st with $s \in S$ and $t \in T$, and z is a sum of elements $(st)u$. Since $(st)u = s(tu)$, these elements all belong to $S(TU)$. Since $S(TU)$ is closed under addition, $z \in S(TU)$. Therefore $(ST)U \subseteq S(TU)$ and the reverse inclusion can be established similarly.

(ii) Let $x, x' \in S + T$. Then $x = s + t$, $x' = s' + t'$ for some s, $s' \in S$ and $t, t' \in T$. Therefore $x - x' = (s - s') + (t - t') \in S + T$ as S and T are additive groups. Further 0 belongs to both S and T and so $0 = 0 + 0 \in S + T$. Hence $S + T$ is an additive subgroup of R.

Now consider ST. We have already seen that it is closed under addition. Furthermore, if $y = \sum s_i t_i \in ST$, then $-y = \sum (-s_i) t_i \in ST$ since $-s_i \in S$. Since ST clearly contains 0, it is therefore an additive subgroup of R.

(iii) We have already seen that ST is a subgroup of R; it remains to show that it is closed under multiplication. However $(\sum_i s_i t_i)(\sum_j s'_j t'_j) = \sum_{i,j} (s_i s'_j)(t_i t'_j)$, since R is commutative; and this element belongs to ST.

2. Homomorphisms

2.4. Definition. A *homomorphism* of a ring R into a ring S is a map $\phi : R \to S$ such that

RH1 $\phi(x + y) = \phi(x) + \phi(y)$

and

RH2 $\phi(xy) = \phi(x)\phi(y)$

for all $x, y \in R$.

Thus by **RH1** such a homomorphism ϕ is in particular a group homomorphism $R^+ \to S^+$, and so by well-known properties of group homomorphisms we have $\phi(0_R) = 0_S$ and $\phi(-r) = -\phi(r)$ for all $r \in R$. Here 0_R denotes the zero of R of course.

As in other branches of mathematics it is customary to prefix the word 'morphism' in various ways to distinguish various important kinds of homomorphism. If R and S are rings, then:

SUBRINGS, HOMOMORPHISMS AND IDEALS

An *epimorphism* $R \to S$ is a surjective homomorphism.
A *monomorphism* $R \to S$ is an injective homomorphism.
An *isomorphism* $R \to S$ is a map which is both an epimorphism and a monomorphism, that is a bijective homomorphism.
An *endomorphism* of a ring R is a homomorphism of R into itself.
An *automorphism* of a ring R is an isomorphism of R into itself.

It is easy to verify that the composition of two homomorphisms is a homomorphism, and that the same is true of any of the 'morphisms' which we have defined above. This follows immediately from the fact that the composition of injective or surjective maps is injective or surjective respectively. Furthermore, if $\phi: R \to S$ is an isomorphism of rings, then the inverse map $\phi^{-1}: S \to R$ (which exists since ϕ is bijective) is also an isomorphism. For, if s, s' are elements of S, then $s = \phi(r)$, $s' = \phi(r')$ for some $r, r' \in R$, and so $\phi^{-1}(ss') = \phi^{-1}(\phi(r)\phi(r')) = \phi^{-1}(\phi(rr')) = rr' = \phi^{-1}(s)\phi^{-1}(s')$. Similarly $\phi^{-1}(s+s') = \phi^{-1}(s) + \phi^{-1}(s')$.

If there exists an isomorphism from R to S, we write $R \cong S$ and say R is (ring) isomorphic to S. The symbol '$\cong$' has the properties of an equivalence relation, that is, (i) $R \cong R$, (ii) $R \cong S \Rightarrow S \cong R$, (iii) $R \cong S$ and $S \cong T \Rightarrow R \cong T$. This follows from what we have said above. Roughly speaking, two rings are isomorphic if one can be obtained from the other simply by renaming the elements but leaving the addition and multiplication tables unchanged, and so isomorphic rings have all ring-theoretic properties in common. The concept of isomorphism now allows us to make precise some vague remarks of Chapter 1. In Ring Example 9 the subsets $\{a\mathbf{1} + b\mathbf{i}\}$, $\{a\mathbf{1} + b\mathbf{j}\}$, etc. are each subrings of the quaternions isomorphic with the (ring of) complex numbers.

Now as we have already remarked, any homomorphism of a ring R to a ring S can be thought of in particular as a homomorphism of R^+ to S^+, and we can quickly obtain some information about such homomorphisms by thinking of them in this way. For example, the *image* $\phi(R)$ or imϕ will be a subgroup of S^+. Also viewed as a group homomorphism ϕ has a *kernel*; this is

$$\{x \in R : \phi(x) = 0_S\}$$

and will often be denoted by kerϕ. We know from elementary group theory that kerϕ is a normal subgroup of R^+ (although the word 'normal' is superfluous here since R^+ is an Abelian group and all its subgroups are normal anyway). By reintroducing the multiplicative structure we can obtain more information about $\operatorname{im}\phi$ and $\ker\phi$. In particular, if x is *any* element of R and $k \in \ker\phi$, then

$$\phi(xk) = \phi(x)\phi(k) = \phi(x)0_S = 0_S.$$

Hence $xk \in \ker\phi$ and similarly $kx \in \ker\phi$.

2.5. Definition. A subset K of a ring R is called an *ideal* of R if K is an additive subgroup of R and whenever $x \in R$, $k \in K$ then $xk \in k$ and $kx \in k$.

The definition can be rephrased in various equivalent ways. Thus in the notation of p. 17, an ideal of R is an additive subgroup K of R satisfying the condition $KR \cup RK \subseteq K$. More explicitly K is an ideal of R if and only if

  (i) $0 \in K$,
  (ii) $k, k' \in K \Rightarrow k - k' \in K$, and
  (iii) $k \in K, x \in R \Rightarrow xk, kx \in k$.

We will write '$K \triangleleft R$' for 'K is an ideal of R'. Examples of ideals will crop up as we go along and for the moment we will content ourselves with remarking that $\{0\}$ and R are always ideals of a ring R.

2.6. Lemma. *Let R and S be rings and $\phi: R \to S$ a homomorphism. Then*:
  (i) $\ker\phi \triangleleft R$. ϕ *is a monomorphism if and only if* $\ker\phi = \{0_R\}$.
  (ii) $\operatorname{im}\phi$ *is a subring of S.*

Proof. (i) We have proved the first statement in our discussion above. Now if ϕ is a monomorphism and $x \in \ker\phi$, then $\phi(x) = 0_S = \phi(0_R)$, hence $x = 0_R$ and $\ker\phi = \{0_R\}$. Conversely, if $\ker\phi = \{0_R\}$, $x, y \in R$ and $\phi(x) = \phi(y)$, then $0_S = \phi(x) - \phi(y) = \phi(x - y)$. Therefore $x - y \in \ker\phi$, $x - y = 0_R$ and $x = y$. Therefore ϕ is a monomorphism in this case.

(ii) We have already seen that $\operatorname{im}\phi$ is an additive subgroup of R. It remains to show that, if $s, s' \in \operatorname{im}\phi$, then $ss' \in \operatorname{im}\phi$. However,

SUBRINGS, HOMOMORPHISMS AND IDEALS 21

by definition there exist $r, r' \in R$ such that $s = \phi(r)$, $s' = \phi(r')$. Therefore $ss' = \phi(r)\phi(r') = \phi(rr') \in \operatorname{im}\phi$.

Having observed that every kernel is an ideal, we are now naturally led to ask whether kernels and ideals are really the same things, that is: Is every ideal of a ring R the kernel of a homomorphism of R into some ring? To answer this it is again instructive to look at the additive group R^+.

Let us then recall the situation for Abelian groups. If A is any Abelian group, and B is a subgroup of A, then a *coset* of B in A is an equivalence class under the equivalence relation $\sim$ on A defined by: $x \sim y \Leftrightarrow x - y \in B$. Since A is Abelian, B is automatically normal in A, and so the usual distinction between right and left cosets disappears. If x is an element of some coset, the elements of that coset are precisely the elements $b + x$ as b runs over the elements of B, and the coset is denoted by $B + x$. The set of all cosets of B in A is denoted by A/B. If we define operations on cosets by

$$(B + x) + (B + y) = B + (x + y)$$

and

$$-(B + x) = B + (-x),$$

then these operations are well-defined; that is, the right-hand side of each of the above equations depends only on the cosets on the left and not on the particular elements x and y chosen to represent the cosets. The operations make A/B into an Abelian group with the coset B as zero element. The map $\nu: x \to B + x$ is a surjective group homomorphism with kernel precisely B; ν is called the *natural homomorphism* of A onto A/B.

Returning to the ring case, we find that we are well on the way to finding a ring homomorphism of R with kernel a given ideal K. We think of R as an additive group, form the set of cosets R/K, make it into an additive group and obtain a group homomorphism $\nu: R \to R/K$ as above. We would like ν to be a ring homomorphism and then all would be well. The main obstacle to this is that R/K is not yet a ring! Can we make it into a ring so as to turn ν into a ring homomorphism? The condition $\nu(x)\nu(y) = \nu(xy)$ would force us to define

$$(K + x)(K + y) = K + xy.$$

So let us try defining multiplication on R/K this way. We must

2

first check that our 'definition' really gives a binary operation on R/K; that is, that the coset on the right depends only on the cosets on the left and not on the elements chosen to represent those cosets. But if $K + x = K + x'$ and $K + y = K + y'$, then $x = x' + k$, $y = y' + l$ for some $k, l \in K$. Then $xy = x'y' + (ky' + x'l + kl)$. Since k and l belong to K and K *is an ideal*, the element in brackets belongs to K. Therefore $K + xy = K + x'y'$, and our definition really does give a binary operation on R/K. Notice that it is exactly the ideal property of K which makes this work.

The rest is completely straightforward. It simply involves checking that R/K satisfies the ring axioms and that ν really is a ring homomorphism. We leave the details to the reader. We may sum up the whole business as follows:

2.7. Lemma. *Let K be an ideal of a ring R, and let R/K be the set of cosets of K in R. Then the definitions*

$$(K + x) + (K + y) = K + (x + y)$$
$$-(K + x) = K + (-x)$$
$$(K + x)(K + y) = K + xy$$

make R/K into a ring. The natural map $\nu: x \to K + x$ is a ring epimorphism with kernel K.

R/K is called the *quotient ring* or *residue class ring* of R by K, and ν is called the *natural homomorphism* from R to R/K. Notice that $(R/K)^+ = R^+/K$.

The fundamental property of the natural homomorphism of a ring R onto a quotient ring R/J is given by the following theorem.

2.8. Theorem. *Let $J \triangleleft R$ and let $\nu: R \to R/J$ be the natural homomorphism. Suppose that $\phi: R \to S$ is a ring homomorphism whose kernel contains J. Then there exists a unique homomorphism $\psi: R/J \to S$ which makes the diagram*

commute. $\ker \psi = \ker \phi / J$.

SUBRINGS, HOMOMORPHISMS AND IDEALS

(The statement that the above diagram commutes means that we get the same result by going from R to S via either of the two possible routes – directly or through R/J. In other words $\phi = \psi \nu$.)

Proof. If the diagram is to commute, then we must have, for an arbitrary element $J + x \in R/J$,

$$\psi(J + x) = \psi\nu(x) = \phi(x), \qquad (*)$$

and so there is only one possible way to define ψ. We must therefore check that defining $\psi(J + x)$ to be $\phi(x)$ does the trick. First, the 'definition' depends only on the coset $J + x$ and not on the representative x. For, if $J + x = J + x'$, then $x - x' \in J$. Hence by assumption $x - x' \in \ker \phi$, and so $\phi(x - x') = 0$. Therefore $\phi(x) = \phi(x')$, and (*) does define a map $\psi: R/J \to S$. If $J + y$ is another element of R/J then $\psi((J + x) + (J + y)) = \psi(J + (x + y)) = \phi(x + y) = \phi(x) + \phi(y) = \psi(J + x) + \psi(J + y)$; similarly ψ preserves multiplication. Therefore ψ is a homomorphism. Finally, from (*) $\psi(J + x) = 0 \Leftrightarrow \phi(x) = 0 \Leftrightarrow x \in \ker \phi$. Therefore $\ker \psi = \ker \phi / J$.

The next three theorems are usually known in some order as the first, second and third isomorphism theorems. They all follow very easily from 2.8.

2.9. Theorem. *If $\phi: R \to S$ is a ring homomorphism, then $R/\ker \phi \cong \operatorname{im} \phi$.*

Proof. In Theorem 2.8 take $J = \ker \phi$. This gives a homomorphism $\mu: R/\ker \phi \to S$ making the diagram

commute, with kernel $\ker \phi / \ker \phi$, the zero subring of $R/\ker \phi$. Therefore by Lemma 2.6 μ is a monomorphism. It follows from the relation $\phi = \mu \nu$ that $\operatorname{im} \mu = \operatorname{im} \phi$. Therefore μ determines an isomorphism of $R/\ker \phi$ onto $\operatorname{im} \phi$.

2.10. Theorem. *Let R be a ring, $J \triangleleft R$ and S a subring of R. Then $S + J$ is a subring of R, $J \triangleleft S + J$, $S \cap J \triangleleft S$ and $S + J/J \cong S/S \cap J$.*

The following diagram may be found helpful in visualizing what this result says.

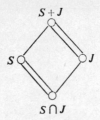

The relation 'is an ideal of' is represented by double lines and the theorem says that the two quotient rings corresponding to the pair of opposite sides with double lines are isomorphic. For this reason the theorem is sometimes known as the 'parallelogram law'.

Proof. We already know by 2.3 that $S+J$ is an additive subgroup of R. Let $s, s' \in S$ and $j, j' \in J$. Then $(s+j)(s'+j') = ss' + (js' + sj' + jj') \in S+J$, since J is an ideal of R. Clearly $J \triangleleft S+J$. Let ν be the natural homomorphism $R \twoheadrightarrow R/J$ and let ν' be the restriction of ν to S. Then ν' is a homomorphism $S \to R/J$. The image of ν' consists of all cosets $s + J$ $(s \in S)$, that is, $\mathrm{im}\,\nu' = S+J/J$. The kernel of ν' consists of all elements in S sent to zero by ν, that is, $\ker \nu = S \cap J$. Hence $S \cap J \triangleleft S$ by 2.6 and $S+J/J \cong S/S \cap J$ by 2.9 above.

2.11. Theorem. *Let R be a ring and let J and K be ideals of R with $J \subseteq K$. Then $K/J \triangleleft R/J$ and $(R/J)/(K/J) \cong R/K$.*

Proof. In 2.8 take ϕ to be the natural homomorphism of R onto R/K. Then $\ker \phi = K$, and we obtain a homomorphism ψ such that the diagram

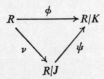

commutes and $\ker \psi = K/J$. Clearly ψ is surjective and so 2.9 applied to ψ gives the result.

SUBRINGS, HOMOMORPHISMS AND IDEALS

There is a further 'isomorphism theorem' which is concerned with the relationship between the ideals, subrings, etc. of $\operatorname{im}\phi$ (where ϕ is a homomorphism of a ring R) and the corresponding objects in R. Before stating it, we recall some set-theoretic notation.

Let A and B be sets and let $f: A \to B$ be a map. Then, if X and Y are subsets of A and B respectively, we define

$$f(X) = \{f(x): x \in X\},$$

and

$$f^{-1}(Y) = \{a \in A: f(a) \in Y\}.$$

The sets $f(X)$ and $f^{-1}(Y)$ are called respectively the *image of X* and the *inverse image of Y*. In this way we obtain a map from the set $\mathscr{P}(X)$ of subsets of X to $\mathscr{P}(Y)$; this map is still denoted by f, although strictly speaking it should be given a different name. Similarly we have a map $f^{-1}: \mathscr{P}(Y) \to \mathscr{P}(X)$. The following statements can easily be verified:

(i) If $Y \subseteq \operatorname{im} f$, then $Y = f(f^{-1}(Y))$.

(ii) If X, X' are subsets of A, and Y, Y' are subsets of B, then $X \subseteq X' \Rightarrow f(X) \subseteq f(X')$ and $Y \subseteq Y' \Rightarrow f^{-1}(Y) \subseteq f^{-1}(Y')$.

Having said all this, we may now state the final 'isomorphism theorem' as follows.

2.12. Theorem. *Let R and S be rings and let $\phi: R \to S$ be a homomorphism with kernel K. Then the maps ϕ and ϕ^{-1} as described above set up an inclusion-preserving bijection between the set of subrings of $\operatorname{im}\phi$ and the set of subrings of R which contain K. In this correspondence, ideals correspond to ideals.*

Proof. Let $T = \operatorname{im}\phi$ and let U be any subring of T. Then we must first show that $\phi^{-1}(U)$ is a subring of R containing K. Now $0 \in U$, and so $\phi^{-1}(U) \geqslant \phi^{-1}(\{0\}) = K$ (using (ii) above). Now let $r, r' \in \phi^{-1}(U)$. Then $\phi(r), \phi(r') \in U$, and so $\phi(r+r') = \phi(r) + \phi(r') \in U$, since U is a subring. Hence $r + r' \in \phi^{-1}(U)$. Similarly $-r$ and rr' belong to $\phi^{-1}(U)$, which is therefore a subring of R. By (i) above, $U = \phi(\phi^{-1}(U))$.

Now let V be any subring of R containing K. We will show that $\phi(V)$ is a subring of T and that $V = \phi^{-1}(\phi(V))$. We shall then have shown that the correspondence set up by ϕ and ϕ^{-1} is bijective between subrings of $T = \text{im}\,\phi$ and subrings of R containing K. Now $\phi(V)$ is just the image of a certain homomorphism of V, namely that obtained by restricting ϕ to V, and is therefore a subring of T by Lemma 2.6. Now let $r \in \phi^{-1}(\phi(V))$. Then $\phi(r) \in \phi(V)$, and so $\phi(r) = \phi(v)$ for some $v \in V$. Hence $\phi(r - v) = 0$, and so $r - v \in K$. Therefore $r = v + k$ for some $k \in K$. However, by assumption $K \subseteq V$, so $k \in V$ and $r \in V$. Hence $\phi^{-1}(\phi(V)) \subseteq V$. The reverse inclusion is obvious, and so $V = \phi^{-1}(\phi(V))$.

The fact that the bijection is inclusion-preserving is just (ii) above. We leave the reader to check that ideals correspond to ideals.

It is instructive to notice what the above correspondence looks like when ϕ is the natural homomorphism ν of a ring R onto a quotient ring R/K. In this case, each subring of $\text{im}\,\nu = R/K$ is a certain set of cosets of K, and ν^{-1} just forms the union of these cosets. In the other direction, each subring of R containing K is a union of cosets of K, and ν replaces the subring by the set of these cosets. Each subring of R/K is the image under ν of a subring V of R containing K, and so has the form V/K; the ideals of R/K are obtained similarly from ideals V of R containing K.

3. Some properties of subrings and ideals

2.13. Lemma. (i) *Let $\{S_\lambda : \lambda \in \Lambda\}$ be any set of subrings (respectively ideals) of a ring R. The $\bigcap_{\lambda \in \Lambda} S_\lambda$ is a subring (resp. ideal) of R.*

(ii) *Let $S_1 \subseteq S_2 \subseteq \ldots$ be an ascending chain of subrings (resp. ideals) of a ring R. Then $\bigcup_{i=1}^\infty S_i$ is a subring (resp. ideal) of R.*

Proof. (i) Since $0 \in S_\lambda$ for all $\lambda \in \Lambda$, $0 \in \bigcap_{\lambda \in \Lambda} S$ which is therefore non-empty. Let $a, b \in \bigcap_{\lambda \in \Lambda} S_\lambda$. Then $a, b \in S_\lambda$ for each $\lambda \in \Lambda$. Hence $a - b$ and ab belong to each S_λ and so also to $\bigcap_{\lambda \in \Lambda} S_\lambda$. Therefore $\bigcap_{\lambda \in \Lambda} S_\lambda$ is a subring. If, furthermore, each S_λ is an ideal of R and $x \in R$, then ax and xa belong to each S_λ and so to $\bigcap_{\lambda \in \Lambda} S_\lambda$. Therefore, in this case, $\bigcap_{\lambda \in \Lambda} S_\lambda$ is an ideal of R.

(ii) Let $S = \bigcup_{i=1}^\infty S_i$. Clearly $0 \in S$. Let $a, b \in S$. Then $a \in S_i$ and $b \in S_j$ for some i and j. Now one of the subrings S_i and S_j contains the other, and so we may choose $l \,(= \max\{i, j\})$ so that a and b both belong to S_l. Then $a - b$ and ab belong to S_l, and so

SUBRINGS, HOMOMORPHISMS AND IDEALS

to S. Therefore S is a subring. The case when the S_i are ideals is left to the reader.

Lemma 2.13 allows us to define the *subring* or the *ideal generated by* a given set of elements. For, if X is a subset of a ring R, then the intersection of all the subrings of R containing X is also a subring of R containing X, and is the smallest such subring. Similar remarks apply to ideals.

2.14. Definition. The *subring generated by a subset* X of a ring R is the smallest subring of R containing X. The *ideal generated by a subset* X is the smallest ideal of R containing X.

It is often useful to have an 'internal' description of the subring or ideal generated by a given set X, that is, one which describes how the elements of the subring or ideal are built up from the elements of X. We now give such a description.

2.15. Lemma. *Let X be a subset of a ring R. Then*:
(i) *The subring of R generated by X consists of all finite sums of elements* $\pm x_1 x_2 \ldots x_n$ ($x_i \in X$, $n = 1, 2, \ldots$).
(ii) *If R is commutative with 1 and $X \neq \varnothing$, then the ideal of R generated by X is RX.*

Proof. (i) Let S be the subring of R generated by X. Since S is a subring of R containing X, S contains all finite products of elements of X and therefore contains the set $\bar{S}$ of all finite sums of elements $\pm x_1 x_2 \ldots x_n$ ($x_i \in X$, $n = 1, 2, \ldots$). On the other hand, $\bar{S}$ contains 0 (this is because we interpret a sum of no terms to be zero – a convenient convention), and is clearly a subring of R containing X. Since S is the smallest such subring, we have $\bar{S} \supseteq S$, and so these two sets are equal.

(ii) Now let R be commutative with 1. We recall that RX denotes the set of all elements of the form $\sum_{i=1}^{n} r_i x_i$ for $r_i \in R$, $x_i \in X$ and $n \geqslant 1$. If $\bar{X}$ denotes the ideal of R generated by X, then every element $r_i x_i$ belongs to $\bar{X}$ and so $RX \subseteq \bar{X}$. On the other hand RX is an ideal of R. For by 2.3 it is an additive subgroup of R, and $R(RX) = (RR)X \subseteq RX$. Since R is commutative, it follows that $RX \triangleleft R$. Furthermore, RX contains X. For if $x \in X$, then $x = 1x \in RX$. The fact that R has a 1 is important here.

It now follows by definition of $\bar{X}$ that $\bar{X} \subseteq RX$ and the two are equal.

The description of the ideal generated by X in more general situations is more complicated (Exercise 11), and we shall not give it here since our main concern is with commutative rings with 1.

We have already defined the sum of two non-empty subsets of a ring (p. 17), and we can extend this definition in the obvious way to finitely many subsets by defining

$$\sum_{i=1}^{n} S_i = S_1 + \cdots + S_n = \{s_1 + \cdots + s_n : s_i \in S_i\}.$$

We then have

2.16. Lemma. *If $J_1, \ldots, J_n$ are ideals of a ring R, then $\sum_{i=1}^{n} J_i$ is an ideal of R.*

Proof. It is clear (cf. Lemma 2.3) that $\sum_{i=1}^{n} J_i$ is an additive subgroup of R. If $j \in \sum_{i=1}^{n} J_i$ and $r \in R$, then $j = \sum_{i=1}^{n} j_i$ with $j_i \in J_i$, and so $rj = \sum_{i=1}^{n} rj_i \in \sum_{i=1}^{n} J_i$. Similarly $jr \in \sum_{i=1}^{n} J_i$, which is therefore an ideal of R.

We shall conclude this chapter by describing the subrings and ideals of $\mathbf{Z}$. It is fairly unusual to be able to describe exactly the set of subrings of a given ring, but in the case of $\mathbf{Z}$ this can be done without too much trouble. We need the following fundamental and familiar property of $\mathbf{Z}$, which we will call the *Euclidean division property*:

If $a, b \in \mathbf{Z}$ and $b \neq 0$, then there exist integers q and r such that

$$a = bq + r \quad \text{and} \quad 0 \leqslant r < |b|.$$

This result is part of our childhood lore of arithmetic, and the main difficulty in offering a proof lies in deciding where to start. We shall be content to draw the following diagram and refer the reader who remains dissatisfied to other books, for example to [6], p. 49, Theorem 12.

SUBRINGS, HOMOMORPHISMS AND IDEALS

2.17. Lemma. *The subrings of $\mathbf{Z}$ are precisely the subrings $n\mathbf{Z} = \{na : a \in \mathbf{Z}\}$ for $0 \leqslant n \in \mathbf{Z}$.*

Proof. It is clear that each of the subsets $n\mathbf{Z}$ is a subring of $\mathbf{Z}$. Now let S be any subring of $\mathbf{Z}$. Then, either $S = \{0\} = 0\mathbf{Z}$, or S contains some non-zero element s. Since S is a subring, S also contains $-s$, and since one of s and $-s$ is positive, it follows that S contains some positive integers. Now any non-empty set of positive integers contains a first, or smallest, member – this is another fundamental property of the set $\mathbf{Z}$ which we are content to quote. Therefore let n be the smallest positive integer belonging to S. Since S is a subring, it therefore contains, with n, the element $na = \pm(n + \cdots + n)$ with $|a|$ terms, a typical element of $n\mathbf{Z}$. Hence $n\mathbf{Z} \subseteq S$. On the other hand, if s is an arbitrary element of S, by the Euclidean division property we can write $s = nq + r$ with q, $r \in \mathbf{Z}$ and $0 \leqslant r < n$. But then $r = s - nq \in S + n\mathbf{Z} \subseteq S$. Therefore, either $r = 0$, or S contains a positive integer smaller than n. Since the second possibility contradicts the choice of n, we have $r = 0$. Hence $s = nq \in n\mathbf{Z}$, and $S \subseteq n\mathbf{Z}$. Therefore $S = n\mathbf{Z}$.

Now clearly $n\mathbf{Z}$ is an ideal of $\mathbf{Z}$; it is in fact the ideal generated by n. Therefore in $\mathbf{Z}$ *every subring is an ideal*. This is a very unusual state of affairs, cf. Exercise 10. The quotient ring $\mathbf{Z}/n\mathbf{Z}$ is the ring $\mathbf{Z}_n$ of *residue classes* mod n already referred to somewhat less precisely in Ring Example 2. Its elements are the cosets $n\mathbf{Z} + m$ as m runs through $\mathbf{Z}$. However, if $n > 0$, then m can be written in the form $qn + r$ with $0 \leqslant r < n$, and so $n\mathbf{Z} + m = n\mathbf{Z} + r$. Hence, if $n > 0$, there are only finitely many distinct cosets, namely $n\mathbf{Z} = n\mathbf{Z} + 0, n\mathbf{Z} + 1, \ldots, n\mathbf{Z} + (n-1)$; these are often written $[0]$, $[1], \ldots, [n-1]$. The operations in $\mathbf{Z}_n$ may be expressed in the form $[i] + [j] = [i+j]$, $[i][j] = [ij]$, $-[i] = [-i]$; the Euclidean division property then enables us to express $[i+j]$, etc. as one of the list $[0], \ldots, [n-1]$ by adding or subtracting a suitable multiple of n.

Exercises for Chapter 2

1. Let R be a ring with 1, and let J be an ideal of R containing 1. Show that $J = R$.

2. List the subrings and ideals of $\mathscr{P}(X)$ (Ring Example 6) in the cases when X has 2 and 3 elements.

3. Let X, Y, Z be non-empty subsets of a ring R. Prove that $X(Y+Z) \subseteq XY + XZ$, and that equality holds if Y and Z both contain 0. Give an example of three subsets of $\mathbf{Z}$ for which equality does not hold.

4. Show that the set-theoretic union of two subrings need not be a subring. In fact, show that if S, S' are subrings of some ring, then $S \cup S'$ is a subring if and only if either $S \subseteq S'$ or $S' \subseteq S$.

5. Show that a field $\mathbf{k}$ has exactly two ideals. More generally, prove that the same is true of $\mathbf{M}_n(\mathbf{k})$.

6. Let $\mathbf{T}_n(\mathbf{k})$ be the set of all $n \times n$ matrices over a field $\mathbf{k}$ with zeros below the diagonal, $\overline{\mathbf{T}}_n(\mathbf{k})$ the subset of such matrices with zeros also on the diagonal, and let $\mathbf{D}_n(\mathbf{k})$ be the set of all $n \times n$ diagonal matrices over $\mathbf{k}$. Show that each of these sets is a ring under the usual operations of matrix addition and multiplication. Prove that $\overline{\mathbf{T}}_n(\mathbf{k}) \triangleleft \mathbf{T}_n(\mathbf{k})$, and that $\mathbf{T}_n(\mathbf{k})/\overline{\mathbf{T}}_n(\mathbf{k}) \cong \mathbf{D}_n(\mathbf{k})$. (*Hint*: construct a homomorphism of $\mathbf{T}_n(\mathbf{k})$ onto $\mathbf{D}_n(\mathbf{k})$ with kernel $\overline{\mathbf{T}}_n(\mathbf{k})$.)

7. Give an example to show that the relation '$\triangleleft$' between subrings of a ring is not transitive. (*Hint*: consider the ring $\mathbf{T}_2(\mathbf{Q})$ defined in the previous question.)

8. Show that any ring homomorphism ϕ can be expressed in the form $\phi = \mu\epsilon$, where ϵ is an epimorphism and μ is a monomorphism.

9. Let ϕ be a homomorphism of an integral domain R into an integral domain S. Show that either $\phi(R) = \{0_S\}$ or $\phi(1_R) = 1_S$.

10. Let R be a ring with 1 in which every subring is an ideal. Show, by considering the subring generated by 1, that either $R = \{0\}$, $R \cong \mathbf{Z}$ or $R \cong \mathbf{Z}_n$. Give an example of a ring without a 1 in which every subring is an ideal.

11. Let R be a ring and X a subset of R. Describe the ideal of R generated by X (a) assuming R has a 1, (b) assuming R is commutative but possibly without a 1, (c) in general.

SUBRINGS, HOMOMORPHISMS AND IDEALS

12. Let R be a commutative ring with 1. Show that R is a field if and only if R has exactly two ideals. An ideal M of a ring R is called *maximal* if there are no ideals J of R satisfying $M \subset J \subset R$. Deduce from Theorem 2.12 that an ideal M of R is maximal if and only if R/M is a field.

13*. Look up Zorn's Lemma if you do not already know it (for example, consult [4], p. 33), and use it to prove that every commutative ring with $1 \neq 0$ has a maximal ideal. Deduce that every such ring can be mapped homomorphically onto a field.

14*. Prove the following extension of Example 12: if R is a commutative ring satisfying $R^2 \neq \{0\}$ and possessing exactly two ideals, then R is a field. Extend the rest of the example correspondingly.

CHAPTER THREE

Construction of new rings

We have already seen in the last chapter how to construct new rings from given ones by forming subrings and quotient rings. In this chapter we discuss three other important constructions for rings – the construction of direct sums, polynomial rings and matrix rings. Such constructions as these are important for several reasons: because they add to our collection of concrete examples, so useful in giving insight into known theorems and in testing the validity of conjectured ones; because it is sometimes possible to prove that certain ring-theoretic properties are inherited by a constructed ring from its component parts, thereby extending a theorem to a larger class of rings; and also because it is often possible to prove theorems which say that certain rings we are interested in may be built up from well-known rings by means of these standard constructions.

1. Direct sums

Let $R_1, \ldots, R_n$ be a finite collection of rings. Let R be the Cartesian product of the sets R_i, and define operations on R 'componentwise', that is

$$(r_1, \ldots, r_n) + (s_1, \ldots, s_n) = (r_1 + s_1, \ldots, r_n + s_n)$$
$$-(r_1, \ldots, r_n) = (-r_1, \ldots, -r_n)$$
$$(r_1, \ldots, r_n)(s_1, \ldots, s_n) = (r_1 s_1, \ldots, r_n s_n).$$

It is easy to check that these operations make R into a ring with $(0, \ldots, 0)$ as the zero element. Notice also that the so-called *co-ordinate projections* $\pi_i : (r_1, \ldots, r_n) \to r_i$ are ring epimorphisms from

R onto R_i. It is sometimes convenient to be able to interpret this direct sum when $n=0$; we do so by defining it as the zero ring $\{0\}$.

3.1. Definition. The ring R defined above is the *external direct sum* of the rings $R_1,\ldots, R_n$, and is denoted by $R_1 \oplus \cdots \oplus R_n$.

Remark. Strictly speaking, there is a certain ambiguity in the definition as given. For suppose that r_1 and s_1 happen to belong not only to the underlying set of R_1 but also to that of R_2. Then the symbol $r+s$ is ambiguous – for it is not clear whether it means addition in R_1 or in R_2. To be precise we ought to mention the operations in each R_i explicitly, writing $r_i +_i s_i$ or the like. However, it seems unlikely that this rather pedantic point will cause confusion, and therefore we will not pursue it further.

It is fruitful to examine the structure of the external direct sum more carefully. For each $i = 1, 2, \ldots, n$ let J_i be the set of all elements $(0, \ldots, 0, r_i, 0, \ldots, 0)$ of R, that is, the set of all n-tuples having zero entries except possibly in the i-th place. The reader may verify that each J_i is an ideal of R, and that the coordinate projection π_i restricted to J_i induces an isomorphism $J_i \to R_i$. (Notice, however, that R_i is not itself a subring of R unless $n=1$.) Also we have $\sum_{i=1}^{n} J_i = R$, and since $\sum_{j \neq i} J_j$ consists of all elements of R whose i-th component is 0, $J_i \cap \sum_{j \neq i} J_j = \{0\}$. These facts help to motivate the following definition.

3.2. Definition. Let R be a ring, and suppose that R has ideals $J_1,\ldots, J_n$ such that

(i) $R = \sum_{i=1}^{n} J_i$, and
(ii) $J_i \cap \sum_{j \neq i} J_j = \{0\}$ for $i=1,\ldots, n$.

Then R is said to be the *internal direct sum* of the ideals J_i. Just as for the external direct sum we write $R = J_1 \oplus \cdots \oplus J_n$, and for $n=0$ we interpret the definition as saying that the zero ring $\{0\}$ is the internal direct sum of an empty collection of ideals. The reason for the use of the same notation for external and internal direct sums will appear shortly. The external direct sum should be thought of as a way of building up more complicated rings from given ones, and the internal direct sum as a way of breaking a given ring down into easier components.

CONSTRUCTION OF NEW RINGS

3.3. Lemma. *Suppose that R is the internal direct sum of its ideals $J_1, \ldots, J_n$. Then every element $r \in R$ has a unique representation in the form*

$$r = r_1 + \cdots + r_n \text{ with } r_i \in J_i.$$

With respect to this representation the operations of R are componentwise.

Proof. Since $R = \sum J_i$, every element of R has at least one representation in the given form. Suppose that $r_1 + \cdots + r_n = r_1' + \cdots + r_n'$ with $r_i, r_i' \in J_i$. Then

$$r_i - r_i' = \sum_{j \neq i} (r_j' - r_j) \in J_i \cap \sum_{j \neq i} J_j = \{0\}.$$

Hence $r_i = r_i'$, and the representation is unique.

Now it is a consequence of the ring axioms that

$$(r_1 + \cdots + r_n) + (s_1 + \cdots + s_n) = (r_1 + s_1) + \cdots + (r_n + s_n),$$

and

$$-(r_1 + \cdots + r_n) = (-r_1) + \cdots + (-r_n),$$

if $r_i, s_i \in J_i$. Since each J_i is an ideal, if $i \neq j$ we have $J_i J_j \subseteq J_i \cap J_j = \{0\}$ by condition (ii) of the definition. Hence $r_i s_j = 0$ if $i \neq j$, and so

$$(r_1 + \cdots + r_n)(s_1 + \cdots + s_n) = r_1 s_1 + \cdots + r_n s_n$$

as claimed.

With the notation of Lemma 3.3, it now follows that the map $\pi_i : r \to r_i$ is a well-defined map of R into J_i. It is called the *projection* of R onto J_i associated with the decomposition $R = J_1 \oplus \cdots \oplus J_n$. Because the operations of R are 'componentwise', it is easy to verify that π_i is an epimorphism.

The relation between internal and external direct sums is now clear. For as we have seen, the external direct sum of a collection $R_1, \ldots, R_n$ of rings is an internal direct sum of ideals $J_i \cong R_i$. On the other hand, a ring R which is an internal direct sum of ideals $J_1, \ldots, J_n$ is isomorphic to the external direct sum of the J_i. The map $\phi : r \to (r_1, \ldots, r_n)$, where r_i is the element of J_i in the unique decomposition of $r = \sum_{i=1}^{n} r_i$, in fact defines an isomorphism of R

with the external direct sum of the J_i. Thus the distinction between internal and external direct sums is essentially a set-theoretic one, and for this reason the same notation is employed for both.

Example. $\mathbf{Z}_6 \cong \mathbf{Z}_2 \oplus \mathbf{Z}_3$.

To see this, let ν_2, ν_3 be the natural homomorphisms of $\mathbf{Z}$ onto $\mathbf{Z}/2\mathbf{Z} = \mathbf{Z}_2$ and $\mathbf{Z}/3\mathbf{Z} = \mathbf{Z}_3$ respectively, and consider the map $\phi : n \to (\nu_2(n), \nu_3(n))$ of $\mathbf{Z}$ into the external direct sum $\mathbf{Z}_2 \oplus \mathbf{Z}_3$. It can easily be verified that this map is a homomorphism and we leave the reader to do so. An integer n belongs to the kernel of ϕ if and only if $(\nu_2(n), \nu_3(n)) = (0, 0)$, that is, if and only if $\nu_2(n) = \nu_3(n) = 0$. Thus the kernel of ϕ is $\ker \nu_2 \cap \ker \nu_3 = 2\mathbf{Z} \cap 3\mathbf{Z} = 6\mathbf{Z}$. Therefore by 2.9 $\mathbf{Z}_6 = \mathbf{Z}/6\mathbf{Z} \cong \mathrm{im}\,\phi$. Now $\mathbf{Z}/6\mathbf{Z}$ has six elements and so $\mathrm{im}\,\phi$ has six elements. Since the number of pairs (a, b) with $a \in \mathbf{Z}_2$ and $b \in \mathbf{Z}_3$ is clearly 6, we have $\mathrm{im}\,\phi = \mathbf{Z}_2 \oplus \mathbf{Z}_3$ and so $\mathbf{Z}_6 \cong \mathbf{Z}_2 \oplus \mathbf{Z}_3$.

If we want to express $\mathbf{Z}_6$ as an internal direct sum $J_2 \oplus J_3$ of ideals isomorphic to $\mathbf{Z}_2$ and $\mathbf{Z}_3$ respectively, we can do so by thinking more carefully about what is going on here. By the proof of 2.9 we have an isomorphism $\psi : \mathbf{Z}_6 \to \mathbf{Z}_2 \oplus \mathbf{Z}_3$ making the diagram

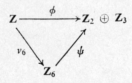

commute. Therefore ψ is actually given by $\psi([a]) = (\nu_2(a), \nu_3(a))$. Now $\mathbf{Z}_2 \oplus \mathbf{Z}_3$ is an internal direct sum $J_2' \oplus J_3'$, where J_2' consists of all elements $(a, 0)$ and J_3' of all elements $(0, b)$. Therefore, using the isomorphism ψ^{-1}, we have $\mathbf{Z}_6 = J_2 \oplus J_3$, where $J_i = \psi^{-1}(J_i')$. But $\psi^{-1}(J_2')$ corresponds to $\ker \nu_3$ and so will consist of $[0]$ and $[3]$, while $\psi^{-1}(J_3')$ corresponds to $\ker \nu_2$ and consists of $[0]$, $[2]$ and $[4]$. The reader will find it a useful exercise to verify directly that $\mathbf{Z}_6$ is the internal direct sum of J_2 and J_3 as defined in this way, and that J_2 and J_3 are isomorphic to $\mathbf{Z}_2$ and $\mathbf{Z}_3$ respectively.

Precisely the same argument shows that $\mathbf{Z}_{rs} \cong \mathbf{Z}_r \oplus \mathbf{Z}_s$ if r and s are integers with no common factors other than ± 1.

CONSTRUCTION OF NEW RINGS

2. Polynomial rings

The reader will almost certainly be familiar with 'polynomials' and might therefore be tempted to disregard this paragraph. Beware! There are two versions of polynomials in circulation, intimately connected and liable to confusion. In this section we give a precise definition of a polynomial ring and show how to recover from our definition the more usual notation for polynomials. We then examine the relationship between our polynomial rings and certain associated rings called rings of polynomial functions, in the hope of dispelling the confusion which is often to be found there.

3.4. Definition. Let R be any ring. The *polynomial ring over* R is the set of all sequences

$$(r_0, r_1, \ldots)$$

with each $r_i \in R$ and all but a finite (but unspecified) number of entries zero; the ring operations are defined as follows:

$$(r_0, r_1, \ldots) + (s_0, s_1, \ldots) = (r_0 + s_0, r_1 + s_1, \ldots)$$
$$-(r_0, r_1, \ldots) = (-r_0, -r_1, \ldots)$$
$$(r_0, r_1, \ldots)(s_0, s_1, \ldots) = (t_0, t_1, \ldots)$$

where $t_i = \sum_{j+k=i} r_j s_k$. Notice that only finitely many terms appear in this sum since if $j + k = i$ then $0 \leqslant j, k \leqslant i$. Strictly speaking of course, the sequences described are maps of a certain kind from the set $\{0, 1, \ldots\}$ to R. However, we prefer to avoid this technically correct notation, fearing it might only conceal the truth from the unfamiliar eye, and leave those well versed in such symbolic punctiliousness to make the translation themselves.

3.5. Theorem. *The above construction yields a ring.*

Proof. Let us temporarily denote the set of sequences described in the above definition by $\bar{R}$. First we must show that the 'operations' given *are* in fact operations on $\bar{R}$. Let $r = (r_0, r_1, \ldots)$ and $s = (s_0, s_1, \ldots)$ be elements of $\bar{R}$. Choose integers $m, n \geqslant 0$ such that $r_i = 0$ for $i > m$ and $s_j = 0$ for $j > n$, and let $l = \max\{m, n\}$. Then

$r_i + s_i = 0$ for $i > l$, and so $r + s \in \bar{R}$. Clearly $-r \in \bar{R}$. Now $(rs)_i = \sum_{j+k=i} r_j s_k$ (where ( )$_i$ denotes the i-th component of the appropriate sequence). Let $i \geqslant m+n+1$. Then in any term $r_j s_k$ with $j+k=i$, either $j > m$ or $k > n$; in the first case $r_j = 0$ and in the second $s_k = 0$, so that in any case $r_j s_k = 0$. Therefore $(rs)_i = 0$ for $i \geqslant m+n+1$, and so $rs \in \bar{R}$.

We must now show that the ring axioms are satisfied. The operation of addition is evidently associative and commutative, the sequence $(0,0,\ldots)$ plays the part of the zero element, and $r + (-r) = 0 \; (= (0,0,\ldots))$ for all $r \in \bar{R}$. Therefore $\bar{R}$ is an Abelian group under $+$. To see that $\bar{R}$ is a multiplicative semigroup we must establish the associativity of multiplication. Let r, s and $t = (t_0, t_1, \ldots)$ be elements of $\bar{R}$. Then

$$((rs)t)_n = \sum_{i+j=n} (rs)_i t_j = \sum_{i+j=n} \left(\sum_{k+l=i} r_k s_l \right) t_j$$

$$= \sum_{k+l+j=n} r_k s_l t_j$$

by the associative and distributive laws in R. Also

$$(r(st))_n = \sum_{k+i=n} r_k (st)_i = \sum_{k+i=n} r_k \left(\sum_{l+j=i} s_l t_j \right)$$

$$= \sum_{k+l+j=n} r_k s_l t_j,$$

and so $(rs)t = r(st)$. It remains only to prove the distributive laws, which we leave as an exercise. The proof is now complete.

We now compare the definition we have just given with the more usual version of 'polynomials'. This can be done most conveniently when R has a 1, and we now assume that to be the case. The reader will easily see that the map $r \to (r,0,0,\ldots)$ is a monomorphism of R into $\bar{R}$, and so the set of all sequences $(r,0,0,\ldots)$ forms a subring of $\bar{R}$ isomorphic to R. We think of R as a subring of $\bar{R}$ by identifying $r \in R$ with the sequence $(r,0,\ldots)$. Now $\bar{R}$ contains the element $(0,1,0,\ldots)$ which we call x. Then from the definition of multiplication we have $x^2 = (0,0,1,0,\ldots)$, $x^3 = (0,0,0,1,0,\ldots)$ and

$$x^n = (\underbrace{0,\ldots,0}_{n},1,0,\ldots) \quad \text{for } n \geqslant 1.$$

CONSTRUCTION OF NEW RINGS

The definition of the operations in R then shows that
$$(r_0, r_1, \ldots, r_n, 0, \ldots) = (r_0, 0, \ldots)(1, 0, \ldots) + (r_1, 0, \ldots)(0, 1, 0, \ldots)$$
$$+ \cdots + (r_n, 0, \ldots)\underbrace{(0, 0, \ldots, 0, 1, 0, \ldots)}_{n}$$
$$= r_0 + r_1 x + \cdots + r_n x^n$$
by our identification. This makes our sequences 'look like' polynomials.

Notation. In view of the above remarks we shall denote the ring R by $R[x]$ and call it the *polynomial ring over R in a single indeterminate x*. The elements of R, identified with elements of $R[x]$ as above, are called the *constant polynomials*. The 'polynomial' notation $r_0 + r_1 x + \cdots + r_n x^n$ will be employed henceforth rather than the 'sequence' notation. The latter is too unwieldy for practical use and has in any case now done its job, which essentially is to emphasize that a polynomial should be thought of as a sequence of coefficients rather than as some kind of 'function'. In times of difficulty we can and should refer to the sequence notation to keep us on the right track.

3.6. Definition. Let R be a ring with 1. Let
$$p = r_0 + r_1 x + \cdots + r_n x^n \in R[x].$$
If $r_n \neq 0$, we say that p has *degree n*, and write $\partial(p) = n$. This associates a degree with every non-zero element of $R[x]$; to complete the definition we conventionally define $\partial(0) = -\infty$. Thus ∂ is a function from $R[x]$ to $\mathbf{Z}_{\geq 0} \cup \{-\infty\}$ where $\mathbf{Z}_{\geq 0} = \{0, 1, 2, \ldots\}$. It is useful to endow the symbol $-\infty$ with some properties by defining
$$n + (-\infty) = (-\infty) + n = -\infty$$
$$(-\infty) + (-\infty) = (-\infty)$$
$$(-\infty) < n$$
for all $n \in \mathbf{Z}_{\geq 0}$. We then have

3.7. Lemma. *If $p, q \in R[x]$, then*:
 (i) $\partial(p + q) \leq \max\{\partial(p), \partial(q)\}$.
 (ii) $\partial(pq) \leq \partial(p) + \partial(q)$.
 (iii) *If R is an integral domain then $\partial(pq) = \partial(p) + \partial(q)$. In this case $R[x]$ is an integral domain also.*

Proof. It is a simple matter to check all the above facts when one or both of p and q is zero. We therefore assume

$$p = r_0 + r_1 + \cdots + r_n x^n \quad (r_n \neq 0)$$
$$q = s_0 + s_1 x + \cdots + s_m x^m \quad (s_m \neq 0)$$

so that $\partial(p) = n$, $\partial(q) = m$. If $l = \max\{m,n\}$, then $p + q = \sum_{i=0}^{l} (r_i + s_i)x^i$, and so $\partial(p+q) \leqslant l$. This proves (i). To see (ii), set $t_i = \sum_{j+k=i} r_j s_k$. Then as in the proof of Theorem 3.5 we have $t_i = 0$ for $i > m + n$, and so $pq = \sum_{i=0}^{m+n} t_i x^i$. Thus $\partial(pq) \leqslant \partial(p) + \partial(q)$. Also $(pq)_{m+n} = r_n s_m$. If R is an integral domain, this element is non-zero, and so $\partial(pq) = m + n$ in this case; in particular, $pq \neq 0$. The commutativity of $R[x]$ follows from that of R, and the element $1 \in R$ is a multiplicative identity in $R[x]$; if R is an integral domain, then $1 \neq 0$, and it follows that $R[x]$ is also an integral domain.

The polynomial ring $R[x]$ is particularly well-behaved in the case when R is a field **k**. Since we shall later be particularly concerned with this situation, we now develop the properties of **k**$[x]$ in more detail. The most fundamental one is the following, which is reminiscent of the Euclidean Division Property for integers (p. 28). We will always use **k** to denote a field.

3.8. Lemma. *Let $a, b \in \mathbf{k}[x]$ and assume $b \neq 0$. Then $\exists\, q, r \in \mathbf{k}[x]$ such that*

$$a = bq + r \quad \text{and} \quad \partial(r) < \partial(b).$$

Moreover q and r are uniquely determined.

Proof. The proof of the existence of q and r is by induction on $\partial(a)$. In order to make the argument absolutely clear we shall write it out with more formality than usual. For $n = 0, 1, \ldots$ let $P(n)$ be the statement that q and r exist provided $\partial(a) < n$. Then $P(0)$ is just the statement that q and r exist when $a = 0$; we can obviously take $q = r = 0$ in that case. We now assume $P(n)$ and deduce $P(n+1)$. To do this, we clearly need only consider the case $\partial(a) = n$. Therefore assume a has the form

$$a = a_0 + a_1 x + \cdots + a_n x^n \quad (a_n \neq 0),$$

and b has the form

$$b = b_0 + b_1 x + \cdots + b_l x^l \quad (b_l \neq 0).$$

CONSTRUCTION OF NEW RINGS

If $n < l$, we simply take $q = 0$, $r = a$. If $n \geq l$, consider $a - a_n b_l^{-1} x^{n-l} b = c$ say. We have arranged matters so that the coefficient of x^n in c is zero. Thus $\partial(c) < n$. Therefore by $P(n)$ we can write
$$c = bq_0 + r \quad \text{with} \quad \partial(r) < \partial(b).$$
Therefore
$$a = b(q_0 + a_n b_l^{-1} x^{n-l}) + r$$
$$= bq + r \quad (\partial(r) < \partial(b)),$$
where $q = q_0 + a_n b_l^{-1} x^{n-l}$. The existence of q and r is now established.

As for the uniqueness, assume $bq + r = bq' + r'$ with $\partial(r)$, $\partial(r') < \partial(b)$. Then $b(q - q') = r' - r$. By Lemma 3.7, $\partial(r' - r) \leq \max\{\partial(r'), \partial(r)\} < \partial(b)$ and $\partial(b(q - q')) = \partial(b) + \partial(q - q')$. Therefore $\partial(b) + \partial(q - q') < \partial(b)$. The only way this can happen is if $\partial(q - q') = -\infty$. Hence $q - q' = 0$, and therefore also $r' - r = 0$, as required.

Notation. Let $c \in \mathbf{k}$ and let $a = a_0 + a_1 x + \cdots + a_n x^n \in \mathbf{k}[x]$. Then we write $a(c)$ for the element $a_0 + a_1 c + \cdots + a_n c^n$ of $\mathbf{k}$. If $a(c) = 0$, we say c is a *root of a*. We leave it as an exercise for the reader to verify that, for a fixed $c \in \mathbf{k}$, the map $a \to a(c)$ is a ring homomorphism of $\mathbf{k}[x]$ into $\mathbf{k}$. (In fact, it is even an epimorphism, since every element of $\mathbf{k}$ is the image of a constant polynomial.) It is this fact which allows us to substitute field elements into 'polynomial identities', as we shall see below.

3.9. Lemma. (*Remainder theorem.*) *Let* $c \in \mathbf{k}$ *and in Theorem* 3.8 *take* $b = x - c$. *Then* $r = a(c)$.

Proof. We have $a = (x - c)q + r$. Here $\partial(r) < \partial(x - c) = 1$, and so r is a constant polynomial. Substitute $x = c$ in this equation, or, in more formal and precise terms, apply the homomorphism $f \to f(c)$. This gives $a(c) = q(c)(c - c) + r = r$ as required.

3.10. Corollary. *If* $a \in \mathbf{k}[x]$, $c \in \mathbf{k}$ *and* c *is a root of* a, *then* $x - c$ *divides* a.

Proof. By Lemma 3.9 $a = (x - c)q + a(c) = (x - c)q$, since $a(c) = 0$ by assumption.

3.11. Theorem. *A polynomial $a \in \mathbf{k}[x]$ of degree $n \geqslant 0$ has at most n distinct roots in* $\mathbf{k}$.

Proof. Let $c_1, \ldots, c_k$ be distinct roots of a in $\mathbf{k}$. We show by induction that $(x - c_1) \ldots (x - c_k)$ divides a. We already know by Corollary 3.10 that $x - c_1$ divides a. Suppose that we have shown that $(x - c_1) \ldots (x - c_i)$ divides a with $i < k$. Then $a = (x - c_1) \ldots (x - c_i)q$ say. Now c_{i+1} is a root of a, and so $0 = a(c_{i+1}) = (c_{i+1} - c_1) \ldots (c_{i+1} - c_i)q(c_{i+1})$. Therefore $q(c_{i+1}) = 0$, c_{i+1} is a root of q, and $x - c_{i+1}$ divides q by 3.10. Therefore $a = (x - c_1) \ldots (x - c_{i+1})q'$ say. In this way we find that $a = (x - c_1) \ldots (x - c_k)\bar{q}$. Therefore $\partial(a) = k + \partial(\bar{q})$. Since $a \neq 0$, $\bar{q} \neq 0$. Therefore $n = \partial(a) \geqslant k$.

The time is now ripe to consider the relationship between the 'formal polynomials' which we have defined and 'polynomial functions'. Let R be a commutative ring with 1. As in Ring Example 8, we can make the set R^R of all functions $R \to R$ into a ring by the pointwise operations given by

$$(f + g)(r) = f(r) + g(r)$$

$$(-f)(r) = -f(r)$$

$$(fg)(r) = f(r)g(r)$$

for $f, g \in R^R$ and $r \in R$.

Let $a = a_0 + a_1 x + \cdots + a_n x^n \in R[x]$. We can associate with a a function $\theta(a) : R \to R$ in the obvious way, that is

$$\theta(a)(r) = a_0 + a_1 r + \cdots + a_n r^n \quad (r \in R).$$

Thus θ is a map from $R[x]$ to R^R. In general θ will not be injective; it can happen that different polynomials determine the same function from R to itself. The simplest example of this is to take R to be the field $\mathbf{Z}_p$ of p elements, where p is a prime, and to consider the polynomial $x^p - x$. The set $\mathbf{Z}_p^*$ of non-zero elements in $\mathbf{Z}_p$ is a multiplicative group of order $p - 1$, and so every element $r \neq 0$ in $\mathbf{Z}_p$ satisfies $r^{p-1} = 1$. Therefore $r^p - r = 0$ for all $r \in \mathbf{Z}_p$ including $r = 0$. This means that the function corresponding to the polynomial $x^p - x$ is the zero function although $x^p - x$ is *not* the zero polynomial.

We shall now show that θ is actually a ring homomorphism from $R[x]$ to R^R. Let $a, b \in R[x]$. Then by allowing some co-

CONSTRUCTION OF NEW RINGS

efficients to be zero we can write $a = \sum_{i=0}^{n} a_i x^i$, $b = \sum_{i=0}^{n} b_i x^i$. Then $a+b = \sum_{i=0}^{n} (a_i + b_i) x^i$. Hence for $r \in R$ we have

$$\theta(a+b)(r) = \sum_{i=0}^{n} (a_i + b_i) r^i = \sum_{i=0}^{n} a_i r^i + \sum_{i=0}^{n} b_i r^i$$
$$= \theta(a)(r) + \theta(b)(r) = (\theta(a) + \theta(b))(r)$$

by definition of addition in R^R. Also

$$ab = \sum_{i=0}^{2n} \left(\sum_{j+k=i} a_j b_k \right) x^i,$$

and so

$$\theta(ab)(r) = \sum_{i=0}^{2n} \left(\sum_{j+k=i} a_j b_k \right) r^i = \sum_{i=0}^{2n} \left(\sum_{j+k=i} a_j r^j \cdot b_k r^k \right)$$
$$= \left(\sum_{j=0}^{n} a_j r^j \right) \left(\sum_{j=0}^{n} b_k r^k \right) = \theta(a)(r) \cdot \theta(b)(r)$$

$= (\theta(a) \theta(b))(r)$ by definition of multiplication in R^R. Thus θ is a ring homomorphism; $\operatorname{im} \theta$ is called the *ring of polynomial functions on R*. A function $f \in R^R$ is thus a polynomial function if and only if there exist $a_0, \ldots, a_n \in R$ such that $f(r) = \sum_{i=0}^{n} a_i r^i$ for all $r \in R$. $\ker \theta$ consists of all elements of $R[x]$ which 'vanish identically' on R, and two polynomials $a, b \in R[x]$ determine the same function on R if and only if $a - b \in \ker \theta$. In the case when R is a field it is easy to give a criterion for the map θ to be injective:

3.12. Theorem. *The map $\theta : \mathbf{k}[x] \to \mathbf{k}^{\mathbf{k}}$ considered above is injective if and only if $\mathbf{k}$ is infinite.*

Proof. First assume that $\mathbf{k}$ is infinite. Let $a \in \ker \theta$. Then $a(r) = 0$ for all $r \in \mathbf{k}$. In other words, every element of $\mathbf{k}$ is a root of a. Now any non-zero element of $\mathbf{k}[x]$ has only finitely many roots by 3.11. Therefore, since a has infinitely many roots, $a = 0$. Therefore $\ker \theta = \{0\}$ and θ is injective.

Now assume that $\mathbf{k}$ is finite, and let $r_1, \ldots, r_n$ be its elements. Then $n \geqslant 1$, and so $(x - r_1) \ldots (x - r_n)$ is a non-zero element of $\mathbf{k}[x]$. This polynomial has every element of $\mathbf{k}$ as a root and so belongs to $\ker \theta$. Hence $\ker \theta \neq \{0\}$ if $\mathbf{k}$ is finite.

A further important property of **k**[x] (where **k** is a field as usual) is that in this ring one can define the concept of 'prime' in a way closely analogous to the usual definition of prime in the ring of integers. One can then prove that every element of **k**[x] can be written as a product of these primes in an essentially unique way. We shall not pursue this topic now as it will be discussed in a more general context later – indeed much of the next chapter is concerned with factorization properties of this type.

3. Matrix rings

If R is any ring, one can define $\mathbf{M}_n(R)$, the set of $n \times n$ matrices with entries in R, just as in the case when R is a field. If addition and multiplication are defined in the usual way, then $\mathbf{M}_n(R)$ becomes a ring – this may be proved exactly as in the case of a field. The principal reason why one meets matrices over a field in the first place is because they arise naturally in studying linear transformations of vector spaces over that field. Since we shall be studying modules over a ring, which in some ways are what you get by replacing the field of a vector space by a more general ring, it will not be surprising if we come across matrices over certain rings later in the text. We shall not require very much information about matrix rings, but the following remarks are of general interest.

Remarks. 1. Suppose that R is a ring with a non-zero multiplication (that is, $R^2 \neq \{0\}$, or equivalently $rs \neq 0$ for some $r, s \in R$). Then

$$\begin{bmatrix} 0 & s \\ 0 & 0 \end{bmatrix} \begin{bmatrix} r & 0 \\ 0 & 0 \end{bmatrix} = \begin{bmatrix} 0 & 0 \\ 0 & 0 \end{bmatrix} \neq \begin{bmatrix} 0 & rs \\ 0 & 0 \end{bmatrix} = \begin{bmatrix} r & 0 \\ 0 & 0 \end{bmatrix} \begin{bmatrix} 0 & s \\ 0 & 0 \end{bmatrix}.$$

Hence $\mathbf{M}_2(R)$, and similarly $\mathbf{M}_n(R)$ for $n > 2$, is non-commutative. In fact, it follows that $\mathbf{M}_n(R)$ is commutative if and only if $n = 1$ and R is commutative.

2. Loosely speaking, $\mathbf{M}_n(R)$ has many subrings and few ideals. The subsets of upper triangular matrices, lower triangular matrices, diagonal matrices, and matrices with zero entries in some specified set of rows or columns, are all subrings. But the interested reader may show that the only ideals of $\mathbf{M}_n(R)$ are the subsets $\mathbf{M}_n(J)$ for $J \triangleleft R$.

CONSTRUCTION OF NEW RINGS

3. A frequently useful, if somewhat pedestrian, way of handling matrices is to single out the n^2 matrices $E_{ij} \in \mathbf{M}_n(R)$, where E_{ij} is the matrix with 1 in the (i, j) position and zeros elsewhere (assuming R has a 1 of course). Then a general element $(r_{ij}) \in \mathbf{M}_n(R)$ can be written uniquely as a linear combination $\sum r_{ij} E_{ij}$. If R is a field $\mathbf{k}$, then $\mathbf{M}_n(\mathbf{k})$ is an n^2-dimensional vector space over $\mathbf{k}$ with the E_{ij} as a basis. The elements E_{ij} multiply according to the rule

$$E_{ij} E_{kl} = \delta_{jk} E_{il},$$

where δ_{jk} is the Kronecker delta. It is easy to check that $\mathbf{M}_n(\mathbf{k})$ is actually an algebra over $\mathbf{k}$ in the sense of the following definition: An *algebra over a field* $\mathbf{k}$ is a set A which is both a ring and a vector space over $\mathbf{k}$ in such a manner that the additive group structures are the same and the axiom

$$\lambda(ab) = (\lambda a) b = a(\lambda b)$$

is satisfied for all $a, b \in A$ and $\lambda \in \mathbf{k}$. This definition, which is important in many contexts, will not be required very much in this book.

4. In the case when R is commutative a determinant function $\mathbf{M}_n(R) \to R$ can be defined exactly as for the case of a field. Most of the usual properties of determinants over a field carry over with unchanged proofs and some of these properties will be required later.

Exercises for Chapter 3

1. Which of the following classes of rings are closed under forming (i) subrings, (ii) quotient rings, (iii) direct sums, (iv) polynomial rings, (v) matrix rings? (a) commutative rings; (b) rings with 1; (c) integral domains; (d) fields. Give a proof or counterexample in each case.

2. Let $S = \{1, 2, \ldots, n\}$ and let R be any ring. Show that R^S, the set of all functions from S to R, made into a ring by the pointwise operations as in Ring Example 8, is isomorphic to the external direct sum $R \oplus \cdots \oplus R$ with n R's.

3. Let X be a finite set with n elements. If E is a subset of X, define a function $\chi_E : X \to \mathbf{Z}_2$ (the characteristic function of E) by
$$\chi_E(x) = 0 \quad (x \notin E)$$
$$\chi_E(x) = 1 \quad (x \in E).$$
Show that $E \to \chi_E$ is an isomorphism from the ring $\mathscr{P}(X)$ (as defined in Ring Example 6) to $\mathbf{Z}_2^X$. Taking into account the previous question, deduce that $\mathscr{P}(X) \cong \mathbf{Z}_2 \oplus \cdots \oplus \mathbf{Z}_2$, n summands.

4. Let R be any ring. Let $\bar{R} = R \oplus \mathbf{Z}$ be the external direct sum of R and $\mathbf{Z}$, considered as an additive group. Define multiplication on $R \oplus \mathbf{Z}$ by $(r,n)(r',n') = (rr' + nr' + n'r, nn')$. Show that this makes $\bar{R}$ into a ring with $(0,1)$ as multiplicative identity, and that the set of all $(r, 0)$ $(r \in R)$ is a subring of $\bar{R}$ isomorphic to R. This allows us to embed an arbitrary ring in a ring with 1.

5. Let R, S, T be rings. Show that $R \oplus (S \oplus T) \cong R \oplus S \oplus T$.

6. Let R be an internal direct sum $R = J_1 \oplus J_2$, and let S be a subring of R containing J_1. Show that $S = J_1 \oplus (S \cap J_2)$. Also show that $R/J_1 \cong J_2$.

7. Show that the rings $\mathbf{Z}$ and $\mathbf{Z} \oplus \mathbf{Z}$ are not isomorphic.

8. Show that the element $(1,0)x^2 - (1,0)x$ of $(\mathbf{Z}_2 \oplus \mathbf{Z}_2)[x]$ has every element of $\mathbf{Z}_2 \oplus \mathbf{Z}_2$ as a root (in the obvious sense). Show that the element $[2]x^2 - [2]x$ of $\mathbf{Z}_4[x]$ has every element of $\mathbf{Z}_4$ as a root. Contrast this with Theorem 3.11.

9. (*Universal property of direct sums.*) Let R_1, R_2 be rings, let $R = R_1 \oplus R_2$ (internal or external), and let $\pi_i : R \to R_i$ be the co-ordinate projections. Show that, given any ring S and

CONSTRUCTION OF NEW RINGS

homomorphisms $\eta_i : S \to R_i$, there exists a unique homomorphism $\eta : S \to R$ which makes the diagrams

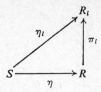

commute. Generalize.

10. Using the idea of the previous exercise, or otherwise, prove that if $J_i \triangleleft R_i$ ($i = 1, 2$), then $R_1 \oplus R_2/J_1 \oplus J_2 \cong R_1/J_1 \oplus R_2/J_2$.

11. Let R be an internal direct sum $R = J_1 \oplus \cdots \oplus J_n$ of ideals J_i. Show that, if $L_i \triangleleft J_i$ for $i = 1, 2, \ldots, n$, then

$$L_1 \oplus L_2 \oplus \cdots \oplus L_n \qquad (*)$$

is an ideal of R.

Now suppose that each J_i is a ring with 1; in other words, that there exists $e_i \in J_i$ such that $e_i x = x e_i = x$ for all $x \in J_i$. Show that every ideal of R has the form (*) in this case. Finally, let J denote the ring obtained from $\mathbf{Z}_2^+$ by defining the product of any two elements to be zero. Find all the ideals of $J \oplus J$, and contrast with the case just considered.

12. (*Universal property of polynomial rings.*) Let R be a commutative ring with 1. Show that given a commutative ring S, a homomorphism $\phi : R \to S$ and an element $a \in S$, there exists a unique homomorphism $\psi : R[x] \to S$ such that (i) $\psi(r) = \phi(r)$ for all $r \in R$, and (ii) $\psi(x) = a$. What happens if R does not have a 1?

13**. Find a polynomial of degree p in $\mathbf{Z}_{p^n}[x]$ (p a prime) which has every element of $\mathbf{Z}_{p^n}$ as a root, and show that there can be no such non-trivial polynomial of smaller degree. Find the smallest degree of a non-trivial polynomial in $\mathbf{Z}_n[x]$ which has every element of $\mathbf{Z}_n$ as a root.

CHAPTER FOUR

Factorization in integral domains

The main result of this chapter is that certain integral domains called principal ideal domains have unique factorization of elements into primes and so behave in that respect just like the integers. We show that a property analogous to the Euclidean division property of $\mathbf{Z}$ is sufficient to ensure that an integral domain is a principal ideal domain.

1. Integral domains

We recall from the definition on p. 12 that an integral domain is a commutative ring with a $1 \neq 0$ and no zero divisors, and that this last requirement implies the cancellation law of multiplication, that is: if $a \neq 0$ and $ax = ay$, then $x = y$. It is worth pointing out, however, that there is no universal convention about the definition of an integral domain; some authors leave out the requirement that $1 \neq 0$, and others drop the requirement of commutativity. The most obvious example of an integral domain is the ring of integers $\mathbf{Z}$ itself. Also any field is an integral domain, and therefore, in particular, $\mathbf{Z}_p$ is an integral domain when p is a prime. However, $\mathbf{Z}_n$ is not an integral domain when n is composite because it has zero divisors; e.g. in $\mathbf{Z}_6$ the elements [3] and [2] are non-zero, but $[3][2] = [6] = [0]$.

Integral domains crop up naturally in several important areas of mathematics, where they frequently occur in the following guises:

(1) Subrings of fields. If $\mathbf{k}$ is a field, it has no zero divisors, because if $a, b \in \mathbf{k}$ with $a \neq 0$ and $ab = 0$, then $b = 1 . b = (a^{-1}a)b = a^{-1}(ab) = a^{-1}0 = 0$. Since $\mathbf{k}$ also has commutative multiplication

and $1 \neq 0$, **k** is an integral domain. It is clear that any subring of **k** which contains the 1 of **k** is also an integral domain. Thus integral domains occur naturally as subrings of fields, and indeed we shall shortly see that every integral domain occurs in this way. Certain subrings of the field **C** of complex numbers figure significantly in Algebraic Number Theory, rings such as the ring of Gaussian integers described in Ring Example 5. Investigation of such rings is often motivated by the integers, involving an attempt to salvage such properties as the existence and uniqueness of factorization into primes.

(2) Polynomial rings. We saw in 3.7 that, if R is an integral domain, so is the polynomial ring $R[x]$; therefore by induction so is the polynomial ring $R[x_1,\ldots,x_n]$ in n indeterminates, which may be defined as $(R[x_1,\ldots,x_{n-1}])[x_n]$. Now the theory of Algebraic Geometry is at heart concerned with the geometrical figures arising as the sets of solutions of polynomial equations in n-dimensional affine and projective space over a field. For example, the unit sphere in three-dimensional Euclidean space can be described as the set of solutions of the equation $x_1^2 + x_2^2 + x_3^2 - 1 = 0$. Not surprisingly, this study involves a careful analysis of the structure of integral domains of the form $k[x_1,\ldots,x_n]$. The machinery of commutative rings needed for the study of algebraic number theory and algebraic geometry is comprehensively treated in [7].

As we have already remarked, every integral domain occurs as a subring of a field. This is the next topic which we want to take up.

4.1. Theorem. *If R is an integral domain, there exists a field **k** which contains a subring isomorphic to R.*

Proof. Our proof is reminiscent of the way one constructs the rational numbers from the integers; as the details are laborious and somewhat space-consuming we give only an outline.

Step 1. On the set of pairs $S = \{(r_1,r_2) : r_1, r_2 \in R, r_2 \neq 0\}$ define a relation $(r_1,r_2) \sim (s_1,s_2)$ if and only if $r_1 s_2 = r_2 s_1$. Show that $\sim$ is an equivalence relation.

Step 2. Denote by $[r_1,r_2]$ the equivalence class of S containing the pair (r_1,r_2), and let **k** be the set of all these equivalence classes.

FACTORIZATION IN INTEGRAL DOMAINS

Bearing in mind the fact that $[r_1, r_2]$ is to represent the fraction r_1/r_2, define operations on the set of equivalence classes by

$$[r_1, r_2] + [s_1, s_2] = [r_1 s_2 + r_2 s_1, r_2 s_2]$$
$$[r_1, r_2][s_1, s_2] = [r_1 s_1, r_2 s_2].$$

Now show that these are well-defined operations on **k**. This involves showing that $r_2 s_2 \neq 0$ (the absence of zero divisors is needed at this point) and that the definitions are independent of class representatives.

Step 3. Verify that **k** satisfies the axioms of a field with these operations, i.e. that **k** and $\mathbf{k} \setminus \{0\}$ are both Abelian groups, the first under addition and the second under multiplication, and that one of the distributive laws holds. The zero element is the class consisting of all pairs $(0, r)$ with $r \neq 0$, and the multiplicative identity is the class of all pairs (r, r) with $r \neq 0$. Also

$$-[r_1, r_2] = [-r_1, r_2],$$

and

$$[r_1, r_2]^{-1} = [r_2, r_1] \quad \text{if } r_1 \neq 0.$$

Step 4. Show that the map $\mu: R \to \mathbf{k}$ defined by $\mu(r) = [r, 1]$ is a monomorphism. Thus $\mu(R)$ is a subring of **k** isomorphic to R.

The field **k** just constructed is usually called the *field of fractions* of the integral domain R. A proof with the details filled in can be found in [6], pp. 135–137.

2. Divisors, units and associates

Our aim is to find analogues of the factorization properties of **Z** in a larger class of rings, in particular in certain integral domains. We have already alluded several times to these 'factorization properties of **Z**'; let us for clarity summarize the facts, with which the reader is no doubt familiar. First of all, an integer p is called a *prime* if (i) $p \neq \pm 1$, and (ii) whenever p is factored as $p = ab$ with $a, b \in \mathbf{Z}$, then one of a and b is ± 1. The unique factorization theorem for **Z** then states:

> Every integer $n \neq 0$ is expressible in the form $\pm 1 \cdot p_1 \ldots p_m$, where $m \geqslant 0$ and the p_i are positive primes. This expression is unique up to the order in which the p_i occur.

It is this theorem which we wish to extend; in fact, our work will imply the above result for **Z** as a special case. We shall work entirely with integral domains throughout this chapter – for, although some of what we say is true more generally, this simplifying assumption may make things clearer.

Notation. We write R^* for the set of non-zero elements of a ring R.

4.2. Definition. If r and s are elements of an integral domain R, then we say r *divides* s (in symbols $r|s$) if there exists an element $t \in R$ such that $s = rt$. In this case r is called a *factor* or a *divisor* of s. Thus the equation $r0 = 0$ means that every element of R is a divisor of zero – in spite of the fact that R has no zero divisors! This apparently ridiculous terminology does not seem to lead to any confusion in practice. Notice, on the other hand, that 0 does not divide any non-zero element of R.

What happens if we examine the problem of factorization in the rational field **Q**? If $r, s \in \mathbf{Q}^*$, then $r = (r/s)s$, and so every element of $\mathbf{Q}^*$ divides every other. Therefore there are no obvious candidates for the primes of **Q**, and certainly no recognizable 'uniqueness of factorization' statements. This difficulty can be avoided, more or less by agreeing to ignore it; to help us to do this we require some more definitions.

4.3. Definitions. (a) Let R be an integral domain. Then a *unit* of R is defined to be a divisor of 1, that is an element u of R such that $uv = 1$ for some $v \in R$.

(b) Again let R be an integral domain. Then two elements r and s of R are said to be *associates* if $r|s$ and $s|r$.

Remarks. 1. Obviously a unit is non-zero. We shall see that the set of all units of an integral domain forms a group under multiplication. For example, the ring **Z** has two units ± 1; these form a cyclic group of order 2 generated by -1. On the other hand, the set of units of **Q** is $\mathbf{Q}^*$, which is as big as it possibly could be – certainly $\mathbf{Q}^*$ is a multiplicative group since **Q** is a field. It follows by examining degrees and using 3.7 that the units of $\mathbf{k}[x]$ are precisely the elements of degree 0, that is the elements of $\mathbf{k}^*$.

2. If $a \in R$ and u is a unit, then there exists v such that $uv = 1$,

FACTORIZATION IN INTEGRAL DOMAINS

and so $a = u(va)$. Thus every unit divides every element of R (just as ± 1 divide every element in $\mathbf{Z}$).

3. Notice that 1 and 2, although not associated in $\mathbf{Z}$, become associated when considered as elements of the larger ring $\mathbf{Q}$. More generally, the elements m and n of $\mathbf{Z}^*$ are only associated in $\mathbf{Z}$ if $m = \pm n$, whereas they are always associated in the ring $\mathbf{Q}$. Thus the concepts of divisibility, unit and associate are relative in that they depend not only on the elements being considered but also on the ring. Sometimes it is necessary to emphasize this by speaking of 'associates in R', and so on.

Notation. We have already defined the product AB of two non-empty subsets A and B of a ring R (p. 17). In the case when $A = \{a\}$ is a set with a single element a, we will write aB instead of $\{a\}B$. It is easy to see that $aR = \{ar : r \in R\}$, although it is not defined in this way. It follows from 2.15 that, if R is an integral domain, then aR (or Ra) is the ideal of R generated by a.

We now prove a portmanteau lemma to put the definitions just given into some perspective.

4.4. Lemma. *Let R be an integral domain. Then*:

(i) $s \mid t$ *if and only if* $sR \supseteq tR$.

(ii) u *is a unit of R if and only if* $uR = R$.

(iii) *The set U of units of R is an (Abelian) group under multiplication, and if $u \in U$ and $v \mid u$, then $v \in U$.*

(iv) *The relation 'is an associate of' is an equivalence relation on R – call it $\sim$ for short. The equivalence class under $\sim$ containing an element a has the form $\{au : u \in U\}$. $a \sim b \Leftrightarrow aR = bR \Leftrightarrow a = bu$ for some unit u.*

(v) *The relation 'divides' is compatible with $\sim$, and when induced to the set of equivalence classes is a partial ordering.*

Proof. (i) If $s \mid t$, then $\exists\, r \in R$ such that $t = sr$. Thus $tR = (sr)R = s(rR) \subseteq sR$. Conversely, assume $tR \subseteq sR$. Then $t = t1 \in tR$, and so $t \in sR$. Hence $t = sr$ for some $r \in R$, and so $s \mid t$.

(ii) Using (i) we have u is a unit $\Leftrightarrow u \mid 1 \Leftrightarrow uR \supseteq 1R = R$. This gives the result.

(iii) If u_1 and u_2 are units, then $\exists\, v_1, v_2 \in R$ such that $u_1 v_1 = u_2 v_2 = 1$. Hence $(u_1 u_2)(v_1 v_2) = (u_1 v_1)(u_2 v_2) = 1$, and so $u_1 u_2 \in U$.

Also $1 \in U$, and v_1 belongs to U and is the multiplicative inverse of u_1. Thus U is a group under multiplication, and clearly it is Abelian since R is commutative. Now suppose that $uw = 1$ and $v|u$. Then $u = vv'$ say, and so $v(v'w) = 1$. Hence v is a unit.

(iv) By definition, $a \sim b$ if and only if $a|b$ and $b|a$. Since $a = 1a$, we certainly have $a \sim a$. $\sim$ is symmetric by its definition. Also $a|b$ and $b|c$ clearly implies $a|c$, and it follows from this that $\sim$ is transitive. Therefore it is an equivalence relation. The rest will follow if we show that $a \sim b \Leftrightarrow b = au$ for some unit u. Suppose then that $a \sim b$, and $a \neq 0$. Then $a|b$ and $b|a$. Hence $b = au$ and $a = bv$ for some $u, v \in R$. Therefore $a = auv$, $uv = 1$ by the cancellation law, and u is a unit. Conversely, assume $b = au$ for some unit u. Then $uv = 1$ for some $v \in R$. Clearly $a|b$, and since we have $a = bv$, also $b|a$. Hence $a \sim b$. The result is true when $a = 0$ since the only element equivalent to 0 is 0 itself.

(v) To say that 'divides' is compatible with $\sim$ means that given two equivalence classes $[a]$ and $[b]$, the definition

$$[a]|[b] \quad \text{if and only if } a|b$$

is independent of the choice of representatives a and b. To see that this is the case, suppose that $[a] = [a']$ and $[b] = [b']$. Then using (iv) and (i), $a|b \Leftrightarrow aR \supseteq bR \Leftrightarrow a'R \supseteq b'R \Leftrightarrow a'|b'$, as required. Now we recall that a partial ordering on a set is a relation ρ on that set which is transitive and antisymmetric, where antisymmetric means that $a\rho b$ and $b\rho a \Leftrightarrow a = b$. Now the transitivity of the relation 'divides' on equivalence classes is obvious from its corresponding property on elements of R. And if $[a]|[b]$ and $[b]|[a]$, then by definition $a|b$ and $b|a$. Hence a and b are associated, and so $[a] = [b]$.

Notation. The set of elements associated to a given element a of an integral domain R will be denoted by $[a]$, as in the above argument. We hope this will not be confused with our use of the same notation to denote the cosets of Z_n (see p. 29).

3. Unique factorization domains

One way of approaching the problem of generalizing a given theorem is to give a name to those rings in which the hoped-for

FACTORIZATION IN INTEGRAL DOMAINS

theorem is true, and then to investigate the class of rings thereby created in order to determine how it relates to other classes of rings. We shall now do this with our 'unique factorization theorem'.

In view of our remarks about units on p. 52, the following is the obvious analogue of the definition of 'prime' for integers. However, it is customary to attach the name 'irreducible' to this concept, and to reserve the name 'prime' for something which is slightly different.

4.5. Definition. Let R be an integral domain. An element $r \in R$ is an *irreducible* of R if (i) r is not a unit, and (ii) whenever $r = ab$ is expressed as a product of two elements a and b of R, one of a and b is a unit (and so the other is an associate of r).

Thus irreducibles are elements whose only factorizations are the trivial ones, effected by means of units. Notice that the equation $0 = 0.0$ means that 0 is not an irreducible.

Remarks. 1. It is easy to see that every associate of an irreducible is irreducible. For, if r is irreducible and $r \sim s$, then $r = us$ for some unit u; obviously s is not a unit. If $s = vt$, then $r = (uv)t$, and so either t or uv is a unit. In the second case v is also a unit by 4.4 (iii).
2. The concept of 'irreducible', like many of the other concepts in this chapter, depends on the ring in which it is interpreted. Thus, although 2 is an irreducible in $\mathbf{Z}$, it is even a unit in the larger ring $\mathbf{Q}$.

4.6. Definition. An integral domain R is called a *unique factorization domain* (UFD for short), or sometimes a *Gaussian domain*, if:
UF1 Every element $r \in R^*$ can be written in the form

$$r = ua_1 \ldots a_n$$

where u is a unit of R, $n \geq 0$ and the a_i are irreducibles of R.
UF2 If $ua_1 \ldots a_n = u'b_1 \ldots b_m$, where u and u' are units and the a_i and b_j are irreducibles, then $n = m$ and $a_i \sim b_{\pi(i)}$ for some permutation π of $\{1, 2, \ldots, n\}$.

Remarks. 1. This resolves the problem which we came across in $\mathbf{Q}$. For there, every non-zero element is a unit. Thus every field is a UFD is a rather trivial way.

2. The existence statement **UF1** is the best analogue we can expect of the corresponding statement for $\mathbf{Z}$, since in general we have no way of picking out particular irreducibles corresponding to the positive primes of $\mathbf{Z}$.

3. Given a factorization $r = ua_1 \ldots a_n$ we can always obtain another in which the a_i are replaced by arbitrary associates $a_i' = u_i a_i$, viz. $r = uu_1^{-1} \ldots u_n^{-1} . a_1' \ldots a_n'$. So the uniqueness statement is also the best we can hope for.

It would be a happy but uninteresting situation if all integral domains were UFD's. But this is a long way from being true – even subrings of the complex numbers need not be UFD's.

Example. Let R denote the subset $\{a+b\sqrt{-5} : a, b \in \mathbf{Z}\}$ of the complex field $\mathbf{C}$. It is not difficult to check that R is a subring of $\mathbf{C}$, and since it contains the 1 of $\mathbf{C}$, that it is an integral domain. First of all we need to determine the units of R. To do this we produce from on high a 'norm' function $n: R \to \mathbf{Z}_{\geq 0}$ defined as follows: for each $\alpha = a + b\sqrt{(-5)} \in R$,

$$n(\alpha) = |\alpha|^2 = a^2 + 5b^2.$$

Here $|\ |$ stands for the usual complex number modulus. This function n has the important property $n(\alpha\beta) = n(\alpha)n(\beta)$, since $|\alpha\beta| = |\alpha||\beta|$. Such a function is said to be *multiplicative*. Now suppose u is some unit of R. Then $uv = 1$ for some $v \in R$, and so $n(u)n(v) = n(1) = 1$. Since $n(u)$ and $n(v)$ are positive integers, we must have $n(u) = n(v) = 1$. But the only integral solutions to the equation $a^2 + 5b^2 = 1$ are $b = 0$ and $a = \pm 1$. These correspond to $u = \pm 1$, and so these are the only units of R.

We now notice that the element $6 = 6 + 0\sqrt{(-5)} \in R$ may be factorized as

$$6 = 2.3 = (1 + \sqrt{-5})(1 - \sqrt{-5}). \qquad (*)$$

Moreover, we claim that each of the four elements $2, 3, 1 + \sqrt{-5}$, $1 - \sqrt{-5}$ is irreducible in R. For example, suppose $2 = \alpha_1 \alpha_2$ with $\alpha_1, \alpha_2 \in R$ and neither α_1 nor α_2 a unit. Using our norm function again we get

$$n(\alpha_1) n(\alpha_2) = n(\alpha_1 \alpha_2) = n(2) = 4.$$

FACTORIZATION IN INTEGRAL DOMAINS

Since $n(\alpha_1)$ and $n(\alpha_2)$ are positive integers, we have $n(\alpha_1) = 1, 2,$ or 4. But, as we saw above, $n(\alpha_1) = 1$ implies that α_1 is a unit, contrary to assumption. And if $n(\alpha_1) = 4$, then $n(\alpha_2) = 1$ and α_2 is a unit. But the equation $a^2 + 5b^2 = 2$ has no solution in integers, and so no element of R has norm 2. Therefore 2 is irreducible in R (it is clearly not a unit as its norm is not 1). Similar arguments establish the irreducibility of 3, $1 + \sqrt{-5}$ and $1 - \sqrt{-5}$.

Since the units of R have norm 1, by the multiplicative property associated elements have equal norm. Therefore 2, which has norm 4, is not associated to either of the elements $1 \pm \sqrt{-5}$, which have norm 6. Therefore **UF2** is not satisfied in R, which is therefore not a UFD.

There is a very important difference between the properties of irreducibles in this ring R and those of prime integers. The reader no doubt realizes that if p is a prime integer, then p has the following property: if $a, b \in \mathbf{Z}$ and $p|ab$, then either $p|a$ or $p|b$. This property is so important that it is usually taken as the definition of 'prime' in general integral domains.

4.7. Definition. An element r of an integral domain R is called *prime* (in R) if (i) r is neither zero nor a unit, and (ii) whenever $a, b \in R$ and $p|ab$, then either $p|a$ or $p|b$.

A glance at the above example reveals that irreducibles are not always prime. For $2|(1 + \sqrt{-5})(1 - \sqrt{-5})$ but 2 does not divide either factor – this can be seen by considering norms.

The following gives an equivalent definition of prime which, though not relevant to our immediate purposes, should be mentioned.

4.8. Lemma. *Let R be an integral domain and let $p \in R^*$. Then p is a prime if and only if R/pR is an integral domain.*

Proof. First assume that p is a prime. Then p is not a unit, and so $p \nmid 1$ and $1 \notin pR$. Therefore $1 + pR \neq pR$ and the additive and multiplicative identities of R/pR are distinct. Clearly R/pR is commutative. Now suppose $(a + pR)(b + pR) = pR$, the zero of R/pR. Then $ab + pR = pR$, $ab \in pR$, and so $p|ab$. Therefore $p|a$ or $p|b$; the first case gives $a + pR = pR$ and the second $b + pR = pR$.

Thus R/pR has no zero divisors, and it is therefore an integral domain.

Now suppose that $p \in R^*$ and R/pR is an integral domain. Then $1 + pR \neq pR$, $p \nmid 1$ and p is not a unit. Furthermore, if $a, b \in R$ and $p|ab$, then $ab + pR = pR$, and since R/pR has no zero divisors, either $a + pR = pR$ or $b + pR = pR$. These two possibilities give $p|a$ and $p|b$ respectively. Therefore p is a prime.

The relationship between primes and irreducibles is of fundamental importance in determining whether a given integral domain is a UFD or not, as we shall now see. One way round the relationship is straightforward:

4.9. Lemma. *Let R be any integral domain. Then in R every prime is irreducible.*

Proof. Let p be a prime of R. Then p is not a unit by definition. Suppose $p = ab$ with $a, b \in R$. Then certainly $p|ab$, so either $p|a$ or $p|b$. In the first case we have $a = pc$ say, with $c \in R$. Then $p = pbc$ and by the cancellation law $bc = 1$. Therefore b is a unit. Similarly, if $p|b$, then a is a unit. Therefore p is indeed irreducible.

4.10. Theorem. *Let R be an integral domain. Then R is a UFD if and only if R satisfies* **UF1** *and*
UF2' *Every irreducible in R is prime.*

Thus in the presence of **UF1**, the conditions **UF2** and **UF2'** are equivalent.

Proof. First assume R is a UFD, and let r be any irreducible of R. Then r is neither a unit nor zero. Suppose that $a, b \in R$ and $r|ab$. Then $rs = ab$ for some $s \in R$. We now use **UF1** to express everything in sight as products of irreducibles; say $s = us_1 \ldots s_l$, $a = va_1 \ldots a_m$, $b = wb_1 \ldots b_n$, where u, v, w are units and the s_i, a_j, b_k are irreducibles. Then from $rs = ab$ we get

$$urs_1 \ldots s_l = (vw) a_1 \ldots a_m b_1 \ldots b_n.$$

Each side of the above equation has the form 'a unit times a product or irreducibles'. Therefore by **UF2**, r is associated either with some a_i of with some b_j. In the first case certainly $r|a_i$ and so $r|a$; in the second $r|b$. Therefore r is prime.

FACTORIZATION IN INTEGRAL DOMAINS

Now we must assume **UF1** and **UF2'** and prove **UF2**. Therefore suppose

$$up_1\ldots p_l = vq_1\ldots q_m, \qquad (*)$$

where $l, m \geqslant 0$, u, v are units and the p's and q's are irreducibles. We have to prove that $l = m$ and that the p's are associated to the q's in some order; we do so by induction on l. If $l = 0$ then we have $u = vq_1\ldots q_m$. If $m > 0$, then since $u|1$, each q_j divides 1, that is each q_j is a unit. This contradicts the definition of irreducibility, and so $m = 0$ if $l = 0$. Now assume that $l > 0$ and that **UF2** holds for equations of the form (*) with fewer than l irreducibles on the left-hand side. Now p_l, being irreducible, is prime since we are assuming **UF2'**. Therefore p_l, which divides the product on the right, divides one of the factors on the right of (*). Now $p_l \nmid v$ (otherwise p_l would be a unit), and so $m > 0$ and by renumbering the q's we may assume that $p_l | q_m$. Since q_m is irreducible its only factors are units and associates of itself; therefore $p_l \sim q_m$ and $p_l = u'q_m$ for some unit u'. Hence substituting in (*) and cancelling the q_m's gives

$$(uu')p_1\ldots p_{l-1} = vq_1\ldots q_{m-1}.$$

Now since **UF2** holds for products with $l - 1$ irreducibles on the left, $l - 1 = m - 1$ and $p_1,\ldots, p_{l-1}$ are associated with $q_1,\ldots, q_{l-1}$ in some order. This gives $l = m$, and since $p_l \sim q_m = q_l$, the result is proved.

It follows from the above two results that in a UFD the concepts of primeness and irreducibility coincide; in particular, this is true in the ring of integers. This explains to some extent why the definition usually given of 'prime' in **Z** is really that of 'irreducible'.

4. Principal ideal domains and Euclidean domains

We now introduce two new classes of rings which turn out to be UFD's.

4.11. Definitions. An ideal J of an integral domain R is called a *principal ideal* of R if J is generated by a single element $a \in R$, that is $J = aR$.

A ring R is called a *principal ideal domain* (PID for short) if R is an integral domain and every ideal of R is principal.

Examples. 1. If R is any integral domain, the ideals $\{0\}$ and R are principal, being generated respectively by 0 and 1.
2. Any field $\mathbf{k}$ is a PID. For it is easy to see (cf. Chapter 2, Exercise 5) that the only ideals of $\mathbf{k}$ are $\{0\}$ and $\mathbf{k}$.
3. $\mathbf{Z}$ is a PID, and if $\mathbf{k}$ is a field, then $\mathbf{k}[x]$ is a PID. We shall prove these statements below. However, in general $R[x]$ is not a PID – see Exercises 8 and 14 at the end of this chapter.

In order to prove **3** above and also to obtain further examples of PID's we introduce a further (and final) class of rings, the *Euclidean domains*. These are obtained by abstracting the 'Euclidean division property' shared by $\mathbf{Z}$ and $\mathbf{k}[x]$ (cf. p. 28 and 3.8).

4.12. Definition. A *Euclidean domain* (ED) is an integral domain R together with a function $\phi: R^* \to \mathbf{Z}_{\geq 0}$ such that
ED1 $a|b \Rightarrow \phi(a) \leq \phi(b)$, and
ED2 given $a \in R$ and $b \in R^*$ there exist $q, r \in R$ such that $a = bq + r$ and either $r = 0$ or $\phi(r) < \phi(b)$. The map ϕ is called a *Euclidean function* on R. There may be many different Euclidean functions which make a given integral domain into a Euclidean domain. As we have seen, $\mathbf{Z}$ and $\mathbf{k}[x]$ are ED's. We shall investigate ED's more closely in §5; however, the reason they interest us is

4.13. Lemma. *Every ED is a PID.*

Proof. The proof is analogous to that of 2.17, where we proved that $\mathbf{Z}$ is a PID (and considerably more). Let R be any ED and suppose $J \triangleleft R$. If $J = \{0\}$, then J is certainly principal. Therefore we may assume $J \neq \{0\}$. Then the set of ϕ-values of non-zero elements of J is a non-empty set of integers ≥ 0 and so contains a smallest member. Let b then be a non-zero element of J whose ϕ-value is as small as possible. We claim that $J = bR$.
Since $b \in J \triangleleft R$, we certainly have $bR \subseteq J$. Conversely, let $a \in J$. Then by **ED2** we have $a = bq + r$ for some $q, r \in R$ with either $r = 0$ or $\phi(r) < \phi(b)$. Now $r = a - bq \in J$. If $r \neq 0$, then $\phi(r) < \phi(b)$, contradicting the choice of b. Therefore $r = 0$, and so $a \in bR$. Hence $J = bR$, as claimed. (Observe that we have not used **ED1** in the proof; however, its value will become clear when we subsequently analyse the units of an ED.)

FACTORIZATION IN INTEGRAL DOMAINS 61

We are now assured of a reasonably plentiful supply of PID's; this makes the next theorem of considerable interest.

4.14. Theorem. *Every PID is a UFD.*

The most convenient way to prove this theorem, taking into account Theorem 4.10, is to prove that every PID satisfies **UF1** and **UF2'**. We deal with these two conditions separately, taking the easier one first.

4.15. Lemma. *Every PID satisfies* **UF2'**.

Proof. Let R be a PID, and let p be an irreducible of R. We have to prove that p is prime. Certainly p is neither zero nor a unit. Let $a, b \in R$, and assume $p|ab$. We shall suppose that $p \nmid a$ and prove that, in that case, $p|b$. Consider the ideal $pR + aR$ of R. As R is a PID this ideal has the form dR for some $d \in R$. Then $d|p$ and $d|a$. Since p is irreducible there are two possibilities: either d is a unit or $d \sim p$. In the second case $p|d$ and so $p|a$, contrary to assumption. Therefore d is a unit and so $pR + aR = R$ by 4.4 (ii). Hence we have $1 = ps + at$ for suitable $r, s \in R$. Therefore $b = psb + tab$; since p divides ab, it divides the right-hand side of this equation, and therefore $p|b$, as required.

The proof of Theorem 4.14 is now completed by

4.16. Lemma. *Every PID satisfies* **UF1**.

Proof. We give a proof by contradiction. Suppose the result is false. Then there exists a PID R and an element $r \in R^*$ which cannot be written in the form 'a unit times a product of irreducibles'. Let us call such elements of R^* 'bad'. The remaining 'good' elements are then those elements of R^* which can be expressed in some way as the product of a unit and a number of irreducibles.

Now our bad element r is in particular neither a unit nor an irreducible itself. Therefore we can write $r = r_1 s_1$, where neither of r_1 and s_1 is a unit and so neither is associated to r. Furthermore, at least one of r_1 and s_1 must be bad, otherwise we could use 'good' factorizations of r_1 and s_1 to give us a good factorization of r. By naming the factors suitably we may suppose that r_1 is bad. Then $r_1|r$ but $r_1 \nsim r$. We now repeat this operation on r_1, and obtain a bad element r_2 such that $r_2|r_1$ and $r_2 \not\sim r_1$. If we continue this process and write $r = r_0$, we obtain an infinite sequence r_0, $r_1, r_2, \ldots$ of bad elements such that $r_{i+1}|r_i$ and $r_{i+1} \not\sim r_i$ for $i = 0$, 1,.... Lemma 4.4 now tells us that the ideals generated by the r_i satisfy

$$Rr_0 \subset Rr_1 \subset Rr_2 \subset \ldots.$$

Let $J = \bigcup_{i=0}^{\infty} Rr_i$. Then by 2.13 $J \triangleleft R$. Therefore, as R is a PID, $J = Rd$ for some $d \in R$. Then $d \in J$, and so d belongs to some Rr_i. Therefore $Rd \subseteq Rr_i \subseteq J = Rd$, and so $J = Rr_i$. But $Rr_i \subset Rr_{i+1} \subseteq J = Rr_i$, and this finally is a contradiction. Therefore no bad elements exist and the lemma is proved.

Remark. We have drawn a veil over the fact that the above argument uses the Axiom of Choice; at the i-th stage of the argument we should have said something like: There exist bad elements dividing r_i but not associated to it; *choose one* and call it r_{i+1}. This would have drawn attention to the fact that our argument requires infinitely many arbitrary choices to be made and thus invokes the Axiom of Choice. The reader who prefers to do so may safely ignore this point; we refer the reader whose curiosity we have succeeded in arousing to [2] or [4] for more details of the Axiom of Choice and related matters.

Summary. The main points of this section can be summarized in the following easily remembered form:

$$\text{ED} \Rightarrow \text{PID} \Rightarrow \text{UFD}.$$

5. More about Euclidean domains

We have already seen that every ED is a PID. The converse of this statement is false; there exist examples of PID's which are not ED's (e.g. the ring of integers of the field $\mathbf{Q}(\sqrt{-19})$) but we shall not attempt to prove this. A discussion of this question and an introductory account of the general problem of factorization can be found in [5]. ED's are usually rather easier to deal with than PID's in general, and this fact justifies us in spending a little more time on them here. The following lemma shows how to recognize the units of an ED.

FACTORIZATION IN INTEGRAL DOMAINS

4.17. Lemma. *Let R be an ED and let $u \in R^*$. Then u is a unit if and only if $\phi(u) = \phi(1)$.*

Proof. If u is a unit then $u|1$ and $1|u$. Hence $\phi(u) = \phi(1)$ by **ED1**.

Conversely, assume $\phi(u) = \phi(1)$, and using **ED2** write $1 = uq + r$, where either $r = 0$ or $\phi(r) < \phi(u) = \phi(1)$. However, since $1|r$, **ED1** gives $\phi(r) \geqslant \phi(1)$ if $r \neq 0$. Hence we must have $r = 0$, and u is a unit.

We have already seen that every ED is a PID; furthermore, we know that an ideal J of an ED is generated by any non-zero element of smallest possible ϕ-value. In an ED there is an explicit way, called the *Euclidean algorithm*, of finding such an element from a given set of generators of J. This we now explain.

Let a, b be elements of an ED R, and assume $b \neq 0$. Use **ED2** to write $a = bq + r$ with $r = 0$ or $\phi(r) < \phi(b)$. We claim that $\{a,b\}$ and $\{b,r\}$ generate the same ideal of R. Let the ideals they generate be J_1 and J_2 respectively. Since $a = bq + r$, we have $a \in J_2$, and therefore $J_1 \subseteq J_2$; on the other hand, $r = a - bq \in J_1$, and so $J_2 \subseteq J_1$. Furthermore, one of two things has happened: either $r = 0$ and the ideal generated by a and b is already generated by b alone, or we have found a new pair $\{b,r\}$ of generators for that ideal such that the ϕ-value of the second generator has been reduced. In the second case we repeat the process. Ultimately, since the ϕ-values are natural numbers and cannot decrease indefinitely, we reduce the second generator to zero. The scheme of calculation is as follows (for convenience writing b_0 and b_1 rather than a and b):

$$b_0 = b_1 q_1 + b_2 \qquad \phi(b_2) < \phi(b_1)$$
$$b_1 = b_2 q_2 + b_3 \qquad \phi(b_3) < \phi(b_2)$$
$$\cdot \ \cdot \ \cdot \ \cdot \ \cdot \ \cdot \ \cdot \ \cdot \ \cdot$$
$$\cdot \ \cdot \ \cdot \ \cdot \ \cdot \ \cdot \ \cdot \ \cdot \ \cdot$$
$$b_{n-1} = b_n q_n + b_{n+1} \qquad \phi(b_{n+1}) < \phi(b_n)$$
$$b_n = b_{n+1} q_{n+1},$$

and by what we have said the pairs $\{b_i, b_{i+1}\}$ all generate the same ideal. Finally we obtain $Rb_0 + Rb_1 = Rb_{n+1}$ which gives us a single generator for the ideal generated by b_0 and b_1. By applying this process several times we can obtain a single generator for an ideal from any given finite set of generators, at each stage replacing some pair of generators by a single generator.

There is another and very important way of looking at the calculations we have just described, and that is in terms of highest common factors.

4.18. Definition. Let R be an integral domain, and let $a_1,\ldots, a_n$ be elements of R. A *highest common factor* (hcf) (sometimes called a greatest common divisor) of $\{a_1,\ldots,a_n\}$ in R is an element $d \in R$ such that

(i) $d|a_i$ for $1 \leqslant i \leqslant n$, and

(ii) if $d' \in R$ and $d'|a_i$ for $1 \leqslant i \leqslant n$, then $d'|d$.

In general, a set of elements of an integral domain may not possess an hcf. However, if d and d^* are hcf's of a set $\{a_1,\ldots,a_n\}$, then by (ii) we have $d|d^*$ and $d^*|d$, that is, $d \sim d^*$. Furthermore, any associate of d has the form du for some unit u and so is also an hcf of $\{a_1,\ldots,a_n\}$. Therefore the set of hcf's of a given set of elements, if non-empty, has the form $[d]$, the set of associates of a particular hcf d. We denote the set of hcf's of a pair of elements a and b by (a,b), bearing in mind that this is a set of elements rather than a single element. Notice, by the way, that the term 'highest' here means highest with respect to the partial ordering of classes of associates introduced in 4.4.

4.19. Lemma. *Any non-empty set $\{a_1,\ldots,a_n\}$ of elements of a PID R possesses hcf's. An element d is an hcf of $\{a_1,\ldots,a_n\}$ if and only if $\sum_{i=1}^n Ra_i = Rd$. Any hcf of $\{a_1,\ldots,a_n\}$ can be expressed in the form $\sum_{i=1}^n r_i a_i$ with $r_i \in R$.*

Proof. Since R is a PID, we know that $\sum_{i=1}^n Ra_i = Rd$ for some $d \in R$. Then, since $a_i \in Rd$, we have $d|a_i$ for $1 \leqslant i \leqslant n$. Furthermore, $d \in \sum_{i=1}^n Ra_i$, and so we have $d = \sum_{i=1}^n r_i a_i$ for some $r_i \in R$. Therefore, if $d' \in R$ and $d'|a_i$ for all i, then $d'|d$. Hence any element d which generates the ideal $\sum_{i=1}^n Ra_i$ is an hcf of $\{a_1,\ldots,a_n\}$. Since any two hcf's are associated, and since associated elements generate the same ideal, the result is proved.

FACTORIZATION IN INTEGRAL DOMAINS

4.20. Corollary. *If R is an ED, then application of the Euclidean algorithm to a pair of elements b_0 and b_1 of R leads to an hcf of b_0 and b_1.*

Notice also that by working from the bottom of the algorithm upwards we can express b_{n+1} explicitly as a linear combination of b_0 and b_1 if we wish.

Worked Examples. 1. *Calculate an hcf of $x^3 + 2x^2 + 4x - 7$ and $x^2 + x - 2$ in $\mathbf{Q}[x]$.*

We first have to subtract multiples of $x^2 + x - 2$ from $x^3 + 2x^2 + 4x - 7$ until we obtain something of degree < 2; the multiple to be subtracted at each stage is determined by considering the terms of highest degree. Thus

$$x^3 + 2x^2 + 4x - 7 = x(x^2 + x - 2) + x^2 + 6x - 7,$$
$$x^2 + 6x - 7 = 1(x^2 + x - 2) + 5x - 5.$$

This gives

$$x^3 + 2x^2 + 4x - 7 = (x^2 + x - 2)(x + 1) + 5x - 5$$

as the first line of the Euclidean algorithm. The next step is

$$x^2 + x - 2 = (5x - 5)(\tfrac{1}{5}x + \tfrac{2}{5}).$$

The remainder is now zero, and so $5x - 5$ (or its associate $x - 1$) is an hcf of the two given polynomials.

2. *Prove that the ring of Gaussian integers is an ED. Find an hcf for $11 + 7i$ and $3 + 7i$ in this ring.*

We recall that the ring R of Gaussian integers is the subring $\{a + bi : a, b \in \mathbf{Z}\}$ of $\mathbf{C}$. If $r = a + bi \in R$, we define $\phi(r) = |r|^2 = a^2 + b^2$, where $|\ |$ is the ordinary complex modulus. Then $\phi(rs) = \phi(r)\phi(s)$, and **ED1** is satisfied.

To verify **ED2**, let $a, b \in R$ with $b \neq 0$, and consider the complex number a/b. Now we can think of the elements of R as the points with integer co-ordinates in the complex plane and divide up the complex plane into squares of side 1 in the obvious way, the vertices of the squares being elements of R. Our complex number a/b falls within or on the boundary of one of these squares; since the diagonal of the square has length $\sqrt{2}$, there is a vertex at a distance $\leq \sqrt{2}/2$ from a/b (see diagram overleaf). Let q be such a vertex. Then $|a/b - q| \leq \sqrt{2}/2 < 1$. Therefore, setting $r = a - bq$, we have $a = bq + r$ and $|r| = |a - bq| = |b||a/b - q| < |b|$. Thus $\phi(r) = |r|^2 < |b|^2 = \phi(b)$. This verifies **ED2**.

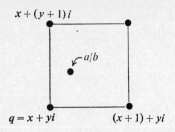

We now use the Euclidean algorithm to find an hcf for $11 + 7i$ and $3 + 7i$. We have

$$11 + 7i/3 + 7i = (11 + 7i)(3 - 7i)/58 = (82 - 56i)/58.$$

The nearest element of R to this is $1 - i$. Therefore

$$11 + 7i = (3 + 7i)(1 - i) + (1 + 3i) \tag{1}$$

is the first line of the algorithm. Next,

$$3 + 7i/1 + 3i = (3 + 7i)(1 - 3i)/10 = (24 - 2i)/10,$$

to which the nearest element of R is 2. Thus the next line of the algorithm is

$$3 + 7i = (1 + 3i).2 + (1 + i). \tag{2}$$

Finally,

$$1 + 3i = (1 + i)(2 + i), \tag{3}$$

and so $1 + i$ is an hcf of $11 + 7i$ and $3 + 7i$. If we want to express $1 + i$ as a linear combination of $11 + 7i$ and $3 + 7i$, we can do so as follows. First (2) gives

$$1 + i = (3 + 7i) - (1 + 3i).2,$$

and substituting for $1 + 3i$ from (1) then gives

$$1 + i = -2(11 + 7i) + (3 - 2i)(3 + 7i).$$

Remark. We have seen that $\mathbf{k}[x]$ is an ED if $\mathbf{k}$ is a field; in particular, $\mathbf{Q}[x]$ is an ED. Also ED $\Rightarrow$ PID $\Rightarrow$ UFD. It is natural to ask where other polynomial rings fit into this scheme – for example, what about $\mathbf{Z}[x]$? In fact (see Exercise 8), $\mathbf{Z}[x]$ is not a PID (and therefore not an ED either). On the other hand, there is an im-

FACTORIZATION IN INTEGRAL DOMAINS

portant theorem due to Gauss which states that *if R is a UFD, so is R[x]*. Therefore $\mathbf{Z}[x]$ is a UFD; so also (by induction on n) is the polynomial ring $\mathbf{k}[x_1,\ldots,x_n]$ $(= (\mathbf{k}[x_1,\ldots,x_{n-1}])[x_n])$. We shall not prove this theorem as it will not be needed in this book. The proof, though fairly lengthy, is not particularly difficult, and the interested reader may find it, for example, in [3], p. 126.

Exercises for Chapter 4

1. Find integers a and b such that $17a + 25b = 1$.

2. In the ring R of Gaussian integers, find an hcf for $a = 5 - i$ and $b = 2 - 2i$ and express it in the form $ra + sb$ with $r, s \in R$. Do the same for $\{2 + i, 3 + 2i\}$.

3. In $\mathbf{Q}[x]$, find an hcf for $x^3 + 2x^2 + 2x + 1$ and $x^2 + x + 2$.

4. In the ring of Gaussian integers, express $1 - 2i$ and $27 + 6i$ as products of primes. Find the units in this ring.

5. Show that in $\mathbf{C}[x]$ every irreducible is linear, and that in $\mathbf{R}[x]$ every irreducible is linear or quadratic.

6. Let $R = \{a + b\sqrt{-5} : a, b \in \mathbf{Z}\}$. Show that the pair of elements $2(1 + \sqrt{-5})$ and $6 = (1 + \sqrt{-5})(1 - \sqrt{-5})$ has no hcf in R. (*Hint*: show that any such hcf would have norm 12.)

7. Show that the rings $\{a + b\sqrt{-2} : a, b \in \mathbf{Z}\}$ and $\{a + b\omega : a, b \in \mathbf{Z}\}$ (where $\omega = e^{2\pi i/3}$), with the usual operations, are Euclidean domains, and find the units in these rings (cf. the proof for the Gaussian integers).

8. Show that $\mathbf{Z}[x]$ is not a PID by considering the ideal J generated by 2 and x.

9. Let R be an integral domain. Prove that, if R is a UFD, then every pair of elements of R has an hcf. Give an example to show that this hcf need not be expressible as a linear combination of the two elements. (Harder) Prove that, if R satisfies

UF1 and every pair of elements of R has an hcf, then R satisfies **UF2'** and hence is a UFD.

10. Let $R \subseteq S$ be PID's, let $a, b \in R$ and let d be an hcf of a and b in R. Prove that d is an hcf of a and b in S.

11. An integral domain R is said to satisfy the *ascending chain condition* on ideals (or to be *Noetherian*, after the distinguished mathematician Emmy Noether (1882–1935)) if given any ascending chain
$$J_1 \subseteq J_2 \subseteq \ldots$$
of ideals of R, there exists an integer n such that $J_n = J_{n+1} = \ldots$. Prove that every PID is Noetherian and that every Noetherian integral domain satisfies **UF1**.

12. Let R be a PID and $p \in R^*$. Show that the following conditions are equivalent: (i) p is prime; (ii) p is irreducible; (iii) pR is a maximal ideal of R (cf. chapter 2, Exercise 12); (iv) R/pR is a field; (v) R/pR is an integral domain.

13. Let R be a PID, S an integral domain, and $\phi: R \to S$ be an epimorphism. Show that either ϕ is an isomorphism or S is a field.

14. Let R be a commutative ring with 1. Show that there exists an epimorphism from $R[x]$ onto R. Deduce that $R[x]$ is a PID if and only if R is a field.

CHAPTER FIVE

Modules

In this chapter we introduce the central concept of the book – that of a module over a ring. In the next few pages we carry out the usual algebraic limbering up process – defining the structure, looking at substructures homomorphisms and quotient structures, and giving plenty of examples. The reader will probably recognize this procedure as an essential preliminary to the challenging heights ahead, and should not be too disappointed if no theorems of great depth and subtlety are proved at this stage.

1. The definition of a module over a ring

A module is a construct of great versatility. It turns up in many seemingly unlikely guises and has the knack of manifesting some of the quintessential features of a wide variety of mathematical structures. It has applications right through modern mathematics, from group theory to topology, and in some areas it is a quite indispensable tool. It provides a language and a way of looking at things which encourages economy of thought and expresses an aesthetically satisfying unity. In case this sounds like an introductory tract for a new idealogy, we must point out at once that such an apparently all-embracing object will probably suffer from some of the defects of such great generality – remoteness from any familiar and significant mathematical situation, a lack of defining detail, the absence of theorems of real depth, and the need for much further effort to obtain useful results in specific situations. We leave the reader to judge for himself.

The concept of a module arises out of an attempt to do classical linear algebra with an arbitrary ring in place of the traditional field. In studying modules one of our aims will be to rescue as much of the classical theory as possible, and at the same time to urge caution by issuing clear warnings where the theory breaks down or needs amendment. Not surprisingly, the increased generality entails a sacrifice. The tidy elegance of theorems about vector spaces has to be abandoned, and our results become more hedged about with 'ifs' and 'buts'. However, the dividend of this painstaking salvage operation will appear in Part III; there we read off some very specific structure theorems about Abelian groups and matrices from our general results about modules.

In writing this text we have assumed the reader is reasonably well versed in the lore of linear algebra, and as we go along we shall emphasize the familiar results of that subject, retrieved, of course, by specializing our ring to a field; thus at two levels in this book the reader's progression will be from the specific to the general and back to the specific again (a sound principle of mathematical education?).

Now down to work! For simplicity we shall consider in this book only modules over rings with 1, and all rings subsequently appearing will be assumed to have a 1 unless the contrary is stated.

5.1. Definition. A *module over a ring R* (or *R-module*) is an Abelian group M (almost invariably written additively) together with a map $(r,m) \to rm$ from $R \times M$ to M satisfying the conditions

M1 $r(m_1 + m_2) = rm_1 + rm_2$,
M2 $(r_1 + r_2)m = r_1 m + r_2 m$,
M3 $(r_1 r_2)m = r_1(r_2 m)$,
M4 $1m = m$,

for all $r, r_1, r_2 \in R$ and all $m, m_1, m_2 \in M$.

It would be more accurate to call what we have just defined a *left R-module*. There is a similar definition of a right R-module in which the elements of R are written on the right. In some situations it is important to have both possibilities together, but as they will not arise in this book we will keep to left modules throughout. Some authors prefer to leave out axiom **M4**, but we shall always include it.

MODULES

Remarks. 1. The first important thing to notice is that the above axioms for a module are precisely the axioms defining a vector space; the only difference is that the so-called scalars are allowed to belong to a ring with 1 instead of being restricted to a field. (If you do not carry the axioms of a vector space in your head, consult any standard reference on linear algebra.)

2. Let M be an R-module. For each $r \in R$ we define a map $\phi(r): M \to M$ by

$$\phi(r)(m) = rm. \qquad (\dagger)$$

Then **M1** says simply that $\phi(r)$ is an endomorphism of the Abelian group M. Thus ϕ is a map from R into $\operatorname{End} M$ (which is a ring under pointwise addition and composition of maps as described in Ring Example 10), and **M3** and **M4** tell us that this map is a ring homomorphism; **M4** simply says that ϕ maps the 1 of R to $1_{\operatorname{End} M}$, the identity map on M.

Conversely, let M be an Abelian group, and suppose we are given a ring homomorphism ϕ from a ring R into $\operatorname{End} M$; suppose further that ϕ sends 1_R to $1_{\operatorname{End} M}$. Then we can use equation ($\dagger$) to make M into an R-module. We leave the reader to check that our hypotheses ensure that axioms **M1**–**M4** are satisfied.

Thus, knowing an R-module is equivalent to knowing a homomorphism of R into the endomorphism ring of some Abelian group. They are simply two ways of viewing or describing the same store of information.

3. Finally, we mention some elementary but very useful consequences of the definition of an R-module M: for all $r \in R$ and $m \in M$ we have

(i) $0_R m = 0_M$,
(ii) $r 0_M = 0_M$, and
(iii) $(-r) m = -(rm) = r(-m)$.

They can be easily verified by careful application of the axioms in the same way as in the proof of Lemma 1.2.

Examples. 1. As we have said, a vector space over a field **k** is precisely the same as a **k**-module.
2. Any Abelian group A can be thought of as a **Z**-module in a natural way. For writing A additively, we have seen on p. 16 how

to define a map $(n,a) \to na$ from $\mathbf{Z} \times A$ to A, and it was pointed out there that this map satisfies conditions corresponding to **M1–M4**.

3. Any ring R (with 1 of course) can be thought of as a module over itself in a natural way. As the additive group we take the additive group R^+ of R, and we define a map $R \times R^+ \to R^+$ by $(r,s) \to rs$, the product of r and s in R. In this case the underlying sets of R and M are the same, but this causes no trouble. Axioms **M1** and **M2** come from the two distributive laws and **M3** from the associative law; **M4** is just the defining property of 1. When we want to emphasize the fact that R is being viewed as a (left) R-module rather than as a ring, we denote it by $_R R$. So now we have three different ways of looking at a ring – as a ring, as an Abelian group, and as a module over itself. In many situations all three are simultaneously important.

4. We have already seen how to view a vector space V over a field $\mathbf{k}$ as a $\mathbf{k}$-module. If we are given a linear transformation of V into itself, we can also make V into a $\mathbf{k}[x]$-module. In order to show how to do this we first recall a few elementary facts about linear transformations. If α and β are linear transformations of V and $\lambda \in \mathbf{k}$, we define maps $\alpha + \beta$, $\alpha\beta$ and $\lambda\alpha: V \to V$ by

$$(\alpha + \beta)(v) = \alpha(v) + \beta(v),$$
$$(\alpha\beta)(v) = \alpha(\beta(v)),$$
$$(\lambda\alpha)(v) = \lambda(\alpha(v)),$$

where v is an arbitrary vector in V. Then it is easy to see that each of these maps is a linear transformation of V and that the operations described made the set End V of all linear transformations of V into a $\mathbf{k}$-algebra (cf. p. 45). Thus, if $\alpha \in$ End V, ι denotes the identity of End V and $f = \sum_{i=0}^{n} a_i x^i \in \mathbf{k}[x]$, then the element $a_0 \iota + a_1 \alpha + \cdots + a_n \alpha^n$ is a well-defined element of End V. We denote it by $f(\alpha)$. Its effect on an arbitrary element $v \in V$ (by definition of the operations in End V) is given by

$$f(\alpha)(v) = a_0 v + a_1 \alpha(v) + \cdots + a_n \alpha^n(v),$$

where $\alpha^n(v) = \alpha(\alpha(\ldots(\alpha(v))\ldots))$ with n α's.

Now let α be a fixed element of End V. We obtain a map $\mathbf{k}[x] \times V \to V$ by defining

$$fv = f(\alpha)(v),$$

MODULES

where $f \in \mathbf{k}[x]$, $v \in V$, and claim that this map makes V into a $\mathbf{k}[x]$-module. We will verify this in detail, since modules of this kind will be very important later in Part III. The simplest verification is to say that by the 'universal property of polynomial rings' described in Chapter 3, Exercise 12, the map $f \to f(\alpha)$ is a ring homomorphism of $\mathbf{k}[x]$ into End V and therefore makes V into a $\mathbf{k}[x]$-module in the way already discussed (see Remark 3 above). However, for those who prefer it we give a verification by direct calculation.

M1 $\quad f(v_1 + v_2) = f(\alpha)(v_1 + v_2)$ (by definition)
$\qquad\qquad\quad = f(\alpha)(v_1) + f(\alpha)(v_2)$ (since $f(\alpha)$ is linear)
$\qquad\qquad\quad = fv_1 + fv_2$ (by definition).

M2 Let $f = \sum_{i=0}^{n} a_i x^i$, $g = \sum_{i=0}^{n} b_i x^i$ be elements of $\mathbf{k}[x]$ (where some of the coefficients may be zero), and let $v \in V$. Then $f + g = \sum_{i=0}^{n} (a_i + b_i) x^i$, and so

$$(f+g)(v) = \sum_{i=0}^{n} (a_i + b_i) \alpha^i(v)$$

$$= \sum_{i=0}^{n} a_i \alpha^i(v) + \sum_{i=0}^{n} b_i \alpha^i(v)$$

$$= fv + gv \quad \text{(by definition)}.$$

M3 With the above notation we have $fg = \sum_{k=0}^{2n} \left(\sum_{i+j=k} a_i b_j \right) x^k$, and therefore

$$(fg)v = \sum_{k=0}^{2n} \left(\sum_{i+j=k} a_i b_j \right) \alpha^k(v) \quad \text{(by definition)}$$

$$= \left(\sum_{i=0}^{n} a_i \alpha^i \right) \left(\sum_{j=0}^{n} b_j \alpha^j(v) \right)$$

$$= f(\alpha)(g(\alpha)(v))$$

$$= f(gv).$$

M4 is immediate.

Notice that the construction depends on specifying a particular α in advance. Different α's will give rise to different maps $\mathbf{k}[x] \times V \to V$ and therefore to different modules. We speak of the $\mathbf{k}[x]$-module constructed as above as the $\mathbf{k}[x]$-*module constructed from V via* α.

5. If A is any Abelian group, then $\operatorname{End} A$ may be given the structure of a ring as in Ring Example 10, and A becomes an $(\operatorname{End} A)$-module if we define $\alpha a = \alpha(a)$ for $\alpha \in \operatorname{End} A$, $a \in A$.

2. Submodules

A submodule of an R-module M will be a subset N of M such that the operations of M, when restricted to N, make N into an R-module. These operations are of two kinds – the Abelian group operations $+$ and $-$, and the operation of multiplying on the left by elements of R. We therefore make the definition:

5.2. Definition. Let M be an R-module. A *submodule* of M is a subset N of M such that
SM1 N is a subgroup of the additive group of M, and
SM2 $rn \in N$ for all $r \in R$ and $n \in N$.

SM2 says that the map $R \times M \to M$ which gives the R-module structure of M maps $R \times N$ into N. It is clear that the module axioms are satisfied, and so N is actually an R-module. The following result is immediate:

5.3. Lemma. *A subset N of an R-module M is a submodule of M if and only if* (i) $0 \in N$, (ii) $n_1, n_2 \in N \Rightarrow n_1 - n_2 \in N$ *and* (iii) $n \in N$, $r \in R \Rightarrow rn \in N$.

Examples. 1. Any R-module M has the submodules M and $\{0\}$.
2. If A is an Abelian group considered as a $\mathbf{Z}$-module as described on p. 45, the submodules of A are precisely the subgroups. This is because, if $n \in \mathbf{Z}$ and $a \in A$, then
$$na = \pm(a + \cdots + a),$$
with $|n|$ terms a, and this belongs automatically to any subgroup containing a.
3. In a vector space over a field $\mathbf{k}$, considered as $\mathbf{k}$-module, the submodules are the subspaces.
4. If R is a commutative ring with 1, then the submodules of $_R R$ are precisely the ideals of R. In the non-commutative case submodules of $_R R$ are called *left ideals*, but we shall not need to refer to these in this book.

MODULES

5. Let V be a vector space over the field $\mathbf{k}$, let $\alpha \in \operatorname{End} V$, and make V into a $\mathbf{k}[x]$-module via α. Suppose that U is a $\mathbf{k}[x]$-submodule of V. Then by considering the effect of constant polynomials we see that U must be a subspace of V. Further since U must be closed under multiplication by x we find

$$\alpha(U) \subseteq U. \qquad (*)$$

Conversely, any subspace U of V satisfying (*) also satisfies $a_0 v + a_1 \alpha(v) + \cdots + a_n \alpha^n(v) \in U$ for any $a_i \in \mathbf{k}$ and $v \in U$, and so is a $\mathbf{k}[x]$-submodule of V. A subspace satisfying (*) is usually called α-*invariant* in linear algebra. Thus the submodules of V, considered as $\mathbf{k}[x]$-module via α, are precisely the α-invariant subspaces of V.

The reader may immediately verify, using Lemma 5.3, that the intersection of any non-empty collection of submodules of an R-module M is itself a submodule of M. This gives us the cue for defining the submodule generated by a subset of M.

5.4. Definition. If X is a subset of an R-module M then the *submodule generated by X* is the smallest submodule of M containing X.

This definition makes sense because the intersection of all the submodules of M containing X is itself such a submodule and is thus the smallest such submodule. In order to describe this submodule more explicitly it is useful to introduce some notation.

Notation. 1. Let M be an R-module. If X is a non-empty subset of M and S is a non-empty subset of R, we denote by SX the set

$$SX = \left\{ \sum_{i=1}^{n} s_i x_i : s_i \in S, x_i \in X, n \geqslant 1 \right\}$$

of all finite sums of elements of the form sx with $s \in S$ and $x \in X$. Thus SX is a subset of M. If M happens to be $_R R$, then X and S are both subsets of R and the product SX defined above is the same as the product of S and X as subsets of R in the sense of p. 17. By its definition SX is closed under addition. If X is an additive subgroup of M, then X contains 0, and so choosing an element s from the non-empty set S we find that SX contains

$s0 = 0$. Also SX contains $-(\sum s_i x_i) = \sum s_i(-x_i)$, since $-x_i \in X$. Thus SX is a subgroup of M in this case. Similarly, if S is a subgroup of R^+, then SX is again a subgroup of M.

Frequently S or X will be sets with a single element, and then we shall write sX instead of $\{s\}X$ and Sx instead of $S\{x\}$. The reader may verify the following statements:

(i) If X is an additive subgroup of M, then $sX = \{sx : x \in X\}$.
(ii) If S is a subgroup of R^+, then $Sx = \{sx : s \in S\}$.
(iii) If $S \triangleleft R$, then SX is a submodule of M.

2. We have already defined the sum of subsets of a ring and the sum of subsets of an R-module may be defined analogously. Thus, if $L_1, \ldots, L_n$ are non-empty subsets of an R-module M, we define

$$\sum_{i=1}^{n} L_i = L_1 + \cdots + L_n = \{l_1 + \cdots + l_n : l_i \in L_i\}.$$

Here we are assuming $n \geqslant 1$. This notation is of particular importance when each L_i is a submodule.

5.5. Lemma. *Let M be an R-module. Then:*

(i) *If $L_1, \ldots, L_n$ are submodules of M ($n \geqslant 1$), then $\sum_{i=1}^{n} L_i$ is a submodule of M.*

(ii) *If X is a non-empty subset of M, then RX is the submodule of M generated by X.*

(iii) *If $X = \{x_1, \ldots, x_n\}$ is a finite non-empty subset of M, then $RX = \sum_{i=1}^{n} Rx_i$.*

Proof. (i) The proof of this will be left to the reader.

(ii) Since R is an ideal of itself, the reader will already have verified that RX is a submodule of M. If $x \in X$, then $x = 1x \in RX$, and so RX contains X. Furthermore, any submodule of M which contains X must contain every element rx ($r \in R$, $x \in X$), and so also every finite sum of such elements. Hence any such submodule contains RX which therefore is the smallest submodule of M containing X.

(iii) We have seen that, if $x \in M$, then $Rx = \{rx : r \in R\}$. Therefore by definition $\sum_{i=1}^{n} Rx_i$ consists of all elements $r_1 x_1 + \cdots + r_n x_n$ with $r_i \in R$. But by definition of RX we can express a typical

MODULES

element of RX in precisely this form, after possibly regrouping terms and using **M2**. Therefore $\sum_{i=1}^{n} Rx_i = RX$, as claimed.

5.6. Definitions. An R-module M is called *finitely-generated* (abbreviated to FG) if M can be generated by some finite set of elements, and *cyclic* if M can be generated by one element.

Thus from 5.5 M is FG if and only if there exist finitely many elements $x_1, \ldots, x_n \in M$ such that each $x \in M$ can be expressed as a 'linear combination' $x = \sum_{i=1}^{n} r_i x_i$ of the x_i with coefficients $r_i \in R$. M is cyclic if and only if $M = Rx$ for some $x \in M$, that is, every element of M has the form rx with $r \in R$ for some fixed $x \in M$.

Examples. 1. Let V be a vector space over a field **k**. Then V is FG as **k**-module if and only if V is finite-dimensional over **k**, and is cyclic if and only if $\dim V = 0$ or 1.
2. Let A be any Abelian group. Then A is FG as **Z**-module if and only if A is FG as a group. A is a cyclic **Z**-module if and only if A is a cyclic group.
3. Let R be a commutative ring with 1 and let M be a submodule of $_R R$. Then $M \triangleleft R$ as we have seen. M is a cyclic submodule of $_R R$ if and only if M is a principal ideal of R. In particular $_R R = R1$ is a cyclic R-module.

We shall have more to say about these concepts later.

3. Homomorphisms and quotient modules

5.7. Definition. Let M and N be R-modules. A *homomorphism* (more precisely, R-module homomorphism or R-homomorphism) from M to N is a map $\theta: M \to N$ such that

MH1 $\theta(m_1 + m_2) = \theta(m_1) + \theta(m_2)$, and
MH2 $\theta(rm) = r\theta(m)$,

for all $m, m_1, m_2 \in M$ and $r \in R$.

Remarks. 1. Notice that M and N are modules *over the same ring*. One cannot sensibly define a module homomorphism from an R-module to an S-module when R and S are different rings.

2. The terms R-module isomorphism, monomorphism, etc. are employed in the usual way to describe homomorphisms which are bijective, injective, etc.

Examples. 1. If M and N are R-modules, the zero map which sends every element of M to 0_N is an R-module homomorphism. The identity map on M is an R-module automorphism.
2. Let A and B be Abelian groups, regarded as $\mathbf{Z}$-modules. Then $\mathbf{Z}$-homomorphisms $A \to B$ are just group homomorphisms.
3. Let V be a vector space over $\mathbf{k}$. Then $\mathbf{k}$-module endomorphisms of V are just linear maps of V into itself. We have already denoted the set of these by End V. A notation such as $\text{End}_{\mathbf{k}} V$ is preferable since it is more explicit and distinguishes $\text{End}_{\mathbf{k}} V$ from $\text{End}_{\mathbf{Z}} V$, the set of all endomorphisms of the additive group of V.
4. Let R be a ring with 1. What are the R-module endomorphisms of ${}_R R$? One might briefly suspect, for no very good reason, that they are just ring endomorphisms. This is not the case. A ring endomorphism $\theta: R \to R$ has to satisfy, for $r, s \in R$,

RH2 $\quad \theta(rs) = \theta(r)\,\theta(s)$,

whereas an R-module endomorphism ϕ of ${}_R R$ has to satisfy

MH2 $\quad \phi(rs) = r\phi(s)$.

For example, taking $R = \mathbf{Z}$, the map $\phi: n \to 2n$ is a $\mathbf{Z}$-module endomorphism but is not a ring endomorphism. For $\phi(1.1) = 2 = 1.\phi(1) \neq \phi(1)\phi(1)$. And if $R = \mathbf{C}$, the map $\theta: x \to \bar{x}$ of complex conjugation is a ring automorphism of $\mathbf{C}$ but is not a $\mathbf{C}$-endomorphism of ${}_{\mathbf{C}}\mathbf{C}$. For $\theta(i.i) = \theta(-1) = -1$, whereas $i\theta(i) = i(-i) = +1$.

The distinction between ring homomorphisms and R-module homomorphisms is a very important one, and care must be taken in situations where confusion might arise.

The development of the elementary theory of R-module homomorphisms now follows the usual pattern, with the customary paraphernalia of kernels, quotient structures, natural homomorphisms, etc. We shall leave much of the detail to the reader, since most of the arguments carry over almost verbatim from

MODULES

Chapter 2, §2 with minor adjustments to account for having left R-action instead of ring multiplication.

First, if M and N are R-modules and $\theta: M \to N$ is an R-homomorphism, then θ is in particular a group homomorphism and so has a kernel

$$\ker \theta = \{m \in M : \theta(m) = 0\}.$$

With the notation above, the following is easily proved:

5.8. Lemma. (i) $\ker \theta$ *is an R-submodule of M.*
(ii) $\operatorname{im} \theta$ *is an R-submodule of N.*

Next one inquires whether every R-submodule K of M is the kernel of some R-homomorphism of M and is rewarded by discovering the quotient module M/K. This consists by definition of all cosets

$$K + m = \{k + m : k \in K\}$$

for all choices of m in M. M/K is already an Abelian group, and we make it into an R-module by defining

$$r(K + m) = K + rm,$$

for each $r \in R$ and each coset $K + m$.

If $K + m = K + m'$, then $m - m' \in K$. Hence $r(m - m') \in K$ since K is a submodule, and so $K + rm = K + rm'$. Therefore the action of R on M/K given above is well-defined. The axioms **M1–M4** are easily verified, and so we have successfully given the set M/K the structure of an R-module; it is called the *quotient module* of M by K. In the special case where M is the R-module $_R R$ and K is an ideal of R, some care is needed to distinguish between the quotient ring R/K and the quotient module of $_R R$ also represented by this symbol. The *natural homomorphism* $\nu: M \to M/K$, given by $\nu(m) = K + m$, is an R-epimorphism with kernel K.

The analogues of theorems 2.8–2.12 now carry over with only the obvious changes. We will content ourselves with stating them.

5.9. Theorem. *Let M and N be R-modules, let K be a submodule of M and let $\nu: M \to M/K$ be the natural homomorphism. Let $\phi: M \to N$ be any R-homomorphism whose kernel contains K.*

Then there exists a unique R-homomorphism $\psi: M/K \to N$ *making the diagram*

commute.

5.10. Theorem. *If M and N are R-modules and $\phi: M \to N$ is an R-homomorphism, then $M/\ker \phi \cong \operatorname{im} \phi$.*

5.11. Theorem. *If K and L are submodules of an R-module M, then $K+L/K \cong L/L \cap K$.*

5.12. Theorem. *If K and L are submodules of an R-module M with $K \subseteq L$, then $(M/K)/(L/K) \cong M/L$.*

5.13. Theorem. *If M and N are R-modules and $\phi: M \to N$ is an R-homomorphism, then ϕ and ϕ^{-1} set up an inclusion-preserving bijection between the set of submodules of M which contain $\ker \phi$ and the set of submodules of $\operatorname{im} \phi$.*

The perspicacious reader no doubt wonders by now why one doesn't find some way of doing all these various isomorphism theorems for groups, rings, vector spaces, modules, etc. at one go. This can in fact be done, although it lies outside the scope of this book and belongs rather to the realms of Universal Algebra (cf. [1]).

4. Direct sums of modules

Direct sums of R-modules (all over the same ring R) are obtained in the usual manner and will play a crucial part in the sequel. For in Part III we aim to decompose a general module of the type we shall consider into a direct sum of submodules which have a particularly easy structure and are in a sense the primitive building blocks of the original structure.

MODULES

5.14. Definition. We say that an R-module M is the *internal direct sum* of submodules $M_1, \ldots, M_n$ if

(i) $M = \sum_{i=1}^n M_i$, and
(ii) $M_i \cap \sum_{j \neq i} M_j = \{0\}$, for $1 \leq i \leq n$.

We write $M = M_1 \oplus \cdots \oplus M_n$, and as usual we regard a zero module as the internal direct sum of an empty collection of submodules.

5.15. Lemma. *If $M_1, \ldots, M_n$ are submodules of M, the following statements are equivalent:*

(i) *M is the direct sum of the M_i.*
(ii) *Each $m \in M$ is uniquely expressible in the form*

$$m = m_1 + \cdots + m_n$$

with $m_i \in M_i$.

Proof. (i) $\Rightarrow$ (ii). By the first requirement of the definition each $m \in M$ is certainly expressible in the form $m = \sum_{i=1}^n m_i$ with $m_i \in M_i$. Suppose we have a second such expression $m = \sum_{i=1}^n \bar{m}_i$ with $\bar{m}_i \in M_i$. Then $m_i - \bar{m}_i = \sum_{j \neq i} (\bar{m}_j - m_j) \in M_i \cap \sum_{j \neq i} M_j = \{0\}$. Hence $m_i = \bar{m}_i$ and the expression is unique.

(ii) $\Rightarrow$ (i). Certainly (ii) implies the truth of the first statement of the definition. If $m \in M_i$, its unique expression is

$$m = 0 + \cdots + 0 + m + 0 + \cdots + 0,$$

where m appears as the i-th term of the sum. But any element in $\sum_{j \neq i} M_j$ has zero in the i-th term of its unique expression. Hence $M_i \cap \sum_{j \neq i} M_j = \{0\}$, and we have established the truth of (i).

The elements m_i occurring in the unique expression described in the second statement of the above lemma are called the *components* of m with respect to the given direct decomposition, and the map $\pi_i : m \to m_i$ is called the *projection* of M on M_i; π_i can be viewed as a map of M into itself, and the reader will have no difficulty in verifying that π_i is an endomorphism of M.

The *external direct sum* of R-modules (each being over the same ring R) is constructed in the obvious way. The underlying set of the external direct sum of R-modules $M_1, \ldots, M_n$ is the set of all n-tuples $(m_1, \ldots, m_n)$ with $m_i \in M_i$. The addition and R-action are co-ordinatewise, that is

$$(m_1, \ldots, m_n) + (\bar{m}_1, \ldots, \bar{m}_n) = (m_1 + \bar{m}_1, \ldots, m_n + \bar{m}_n),$$
$$r(m_1, \ldots, m_n) = (rm_1, \ldots, rm_n).$$

It is trivial to check that this gives an R-module, denoted as usual by $M_1 \oplus \cdots \oplus M_n$. The set of all n-tuples which have zeros everywhere except possibly in the i-th component forms a submodule $\bar{M}_i$ isomorphic to M_i, and M is the internal direct sum of the $\bar{M}_i$, just as in the ring case. Furthermore, any internal direct sum of submodules is isomorphic to the external direct sum of those modules.

Exercises for Chapter 5

(R will denote a commutative ring with 1 unless otherwise stated.)

1. Let R be the subring $\{a + b\sqrt{2} : a, b \in \mathbf{Z}\}$ of $\mathbf{C}$. R can be thought of either as a $\mathbf{Z}$-module or as an R-module (see Examples 2 and 3 on pp. 71–72). Show that the map $a + b\sqrt{2} \to a + b$ is a $\mathbf{Z}$-endomorphism of R but is not an R-endomorphism nor a ring endomorphism. Show that, as $\mathbf{Z}$-module, $R \cong {}_\mathbf{Z}\mathbf{Z} \oplus {}_\mathbf{Z}\mathbf{Z}$.

2. Let V be a vector space over a field $\mathbf{k}$ with basis $\{v_1, v_2\}$, and let $\alpha : V \to V$ be the map defined by $\alpha(\lambda_1 v_1 + \lambda_2 v_2) = \lambda_2 v_1 + \lambda_1 v_2$ for all $\lambda_1, \lambda_2 \in \mathbf{k}$. Show $\alpha \in \text{End}_\mathbf{k} V$, and describe (by finding bases) all the submodules of V considered as a $\mathbf{k}[x]$-module via α. Contrast this with the case where V is considered as a $\mathbf{k}$-module. (*Caution*: $\mathbf{k}$ may have characteristic 2.)

3. Show that the subset $2\mathbf{Z}$ of the $\mathbf{Z}$-module $\mathbf{Z}$ is a submodule. Show further that $2\mathbf{Z}$ is module isomorphic but not ring isomorphic with $\mathbf{Z}$.

4. To generalize the previous exercise, let R be an integral domain and x a non-zero element of R. Show that $R \cong Rx$ as R-modules. Show that $R \cong Rx$ as rings if and only if x is a unit.

MODULES

5. Show that $R[x]$ is FG as R-module if and only if $R = 0$. Show that $\mathbf{Q}$ is not FG as a $\mathbf{Z}$-module.

6. Find a natural way of making $\mathbf{M}_n(R)$ into an R-module, and show that then $\mathbf{M}_n(R) \cong {}_RR \oplus \cdots \oplus {}_RR$, the external direct sum of ${}_RR$ with itself n^2 times.

7. Let M be an R-module and r a fixed element of R. Show that the map $m \to rm$ is an R-endomorphism of M. Denoting the kernel of this endomorphism by M_r, prove that $M/M_r \cong rM$. If $M = M_1 \oplus M_2$, the internal direct sum of submodules, prove that $rM = rM_1 \oplus rM_2$ and $M_r = (M_1)_r \oplus (M_2)_r$.

8. Let M be an R-module and suppose $M = L \oplus N$. If $L = L_1 \oplus \cdots \oplus L_k$ and $N = N_1 \oplus \cdots \oplus N_l$, prove that $M = L_1 \oplus \cdots \oplus L_k \oplus N_1 \oplus \cdots \oplus N_l$ (all direct sums are internal). Generalize this result.

9. Suppose that M_1, M_2, N_1, N_2 are R-modules and that $M_i \cong N_i$ for $i = 1, 2$. Prove that $M_1 \oplus M_2 \cong N_1 \oplus N_2$. Is the converse true?

10. Let M be an R-module and set $J = \{r \in R : rM = \{0\}\}$. Show that $J \triangleleft R$ and that M may be made into a module for R/J in a natural way.

11. Let V_1 and V_2 be vector spaces over a field $\mathbf{k}$, made into $\mathbf{k}[x]$-modules via elements $\alpha_i \in \mathrm{End}_{\mathbf{k}} V_i$ ($i = 1, 2$). Show that $V_1 \cong V_2$ as $\mathbf{k}[x]$-modules if and only if $\alpha_1 = \gamma^{-1}\alpha_2\gamma$ for some vector space isomorphism $\gamma : V_1 \to V_2$.

12. Let M be any R-module and let $E = \mathrm{End}_R M$ be the set of R-endomorphisms of M. Show that the definitions
$$(\eta_1 + \eta_2)(m) = \eta_1(m) + \eta_2(m)$$
$$(\eta_1 \eta_2)(m) = \eta_1(\eta_2(m))$$
($\eta_1, \eta_2 \in E$, $m \in M$) make E into a ring. Show that M can be regarded as an E-module and that every element of R determines an E-endomorphism of M.

13. Let $M = M_1 \oplus \cdots \oplus M_n$ be an internal direct sum of submodules and let π_i be the associated projection on M_i. Show that (i) $\sum_{i=1}^n \pi_i = 1$, (ii) $\pi_i^2 = \pi_i$ and (iii) $\pi_i \pi_j = 0$ if $i \neq j$, where 1 and 0 denote the identity and zero endomorphism of M respectively.

Suppose that $\pi_1, \ldots, \pi_n$ are endomorphisms of an arbitrary module M satisfying (i)–(iii) above, and let $M_i = \operatorname{im} \pi_i$. Show that $M = M_1 \oplus \cdots \oplus M_n$ and that the π_i are the associated projections.

14*. If M and N are R-modules, let $\operatorname{Hom}_R(M, N)$ denote the set of R-homomorphisms from M to N. Show that the point-wise definition of addition makes $\operatorname{Hom}_R(M, N)$ into an Abelian group.

Let $M = M_1 \oplus M_2$ be an internal direct sum of submodules and let π_1, π_2 be the associated projections. Prove that, if $\phi \in \operatorname{End}_R M$ then $\phi = \sum_{i,j=1}^2 \pi_i \phi \pi_j$. Let $\phi_{ij} = \pi_i \phi \pi_j |_{M_j}$. We then have a map

$$\phi \to \begin{bmatrix} \phi_{11} & \phi_{12} \\ \phi_{21} & \phi_{22} \end{bmatrix},$$

where $\phi_{ij} \in \operatorname{Hom}_R(M_j, M_i)$. Show that the usual operations of matrix addition and multiplication make the set of all such matrices into a ring and that the above map is then a ring isomorphism.

Interpret this in the case when M is a 2-dimensional vector space over a field. Generalize to the case of n summands.

Determine $\operatorname{End}_\mathbf{Z} A$ when $A = \mathbf{Z}_2, \mathbf{Z}_3, \mathbf{Z}_3 \oplus \mathbf{Z}_3, \mathbf{Z}_3 \oplus \mathbf{Z}_2$.

CHAPTER SIX

Some special classes of modules

Modules are far too varied and complicated to study in the large. But by restricting their nature in different ways we can focus on smaller areas of the subject and describe more clearly what we see. In this chapter we pick out for special attention several characteristics of modules which make their study more tractable. Since the aim of this book is not to provide a comprehensive introductory treatise on module theory, but rather to illustrate its value in one small corner of modern algebra, we have allowed our selection to be governed entirely by the needs of what is to follow.

1. More on finitely-generated modules

We have already met finitely-generated modules in Chapter 5, §2. We recall that an R-module M is FG if and only if there exist finitely many elements $m_1, \ldots, m_n$ in M such that every element $m \in M$ can be expressed (possibly in several ways) as a linear combination

$$m = r_1 m_1 + \cdots + r_n m_n,$$

with coefficients $r_i \in R$. It is useful to know how the property of being FG behaves under the operations on modules which were introduced in the last chapter.

6.1. Lemma. *Let M be an R-module. Then:*

 (i) *If M is the sum of finitely many FG submodules, then M is FG.*

 (ii) *If M can be generated by s elements and N is a submodule of M, then M/N can be generated by s elements.*

(iii) *If* $M = M_1 \oplus M_2$ *and* M *can be generated by* s *elements, then* M_1 *can be generated by* s *elements.*

Proof. (i) is clear.

(ii) By assumption there exist s elements $m_1, \ldots, m_s \in M$ such that any element $m \in M$ has the form $m = \sum_{i=1}^{s} r_i m_i$ with $r_i \in R$. Then $N + m = \sum_{i=1}^{s} r_i (N + m_i)$, which shows that the s elements $N + m_1, \ldots, N + m_s$ generate M/N.

(iii) By 5.11 we have $M/M_2 = M_1 \oplus M_2/M_2 \cong M_1/M_1 \cap M_2 = M_1/\{0\} \cong M_1$. Now by (ii) M/M_2 can be generated by s elements. Hence M_1 can be generated by s elements.

However, although every direct summand of a FG module is FG, an FG module can have submodules which are not FG. Contrast this with the vector space situation, where every subspace of a finite-dimensional space is finite-dimensional.

Example. Let R be the ring of all maps $\mathbf{R} \to \mathbf{R}$ (where the ring operations are pointwise as in Ring Example 8). Then R is a commutative ring with multiplicative identity, viz. the map which sends every element of $\mathbf{R}$ to 1. Hence $M = {}_R R$ is a cyclic R-module, and is therefore certainly FG.

Let N be the set of all $f \in R$ which vanish outside some finite interval. Thus $f \in N$ if and only if there exists an integer $n \geq 0$, depending on f of course, such that $f(x) = 0$ whenever $|x| > n$. Clearly, if $f, g \in N$, then $f - g \in N$, and if $h \in R$, then $hf \in N$, since hf vanishes whenever f does. Furthermore, the zero function 0 belongs to N. Thus N is a submodule of M.

Let $\{f_1, \ldots, f_k\}$ be a finite set of functions in N. Then for each i there exists an integer n_i such that $f_i(x) = 0$ whenever $|x| > n_i$. If $n = \max n_i$, then each of $f_1, \ldots, f_k$ vanishes outside $[-n, n]$. Hence any linear combination $\sum h_i f_i$ ($h_i \in R$) vanishes outside $[-n, n]$. Therefore $f_1, \ldots, f_k$ do not generate N; for example, the function which takes the value 1 at every point of $[-(n+1), (n+1)]$ and is zero outside that interval belongs to N but is not a linear combination of the f_i. We have thus proved that N is not FG.

It is not difficult here to replace the ring R by much smaller rings, for example the ring of infinitely differentiable functions from $\mathbf{R}$ to $\mathbf{R}$.

SOME SPECIAL CLASSES OF MODULES

2. Torsion modules

6.2. Definition. An element m of an R-module M is called a *torsion element* if there exists a *non-zero* element $r \in R$ such that $rm = 0$. A *torsion module* is one all of whose elements are torsion elements. At the other extreme a module in which there are no non-zero torsion elements is called *torsion-free*. An element which is not a torsion element is called a *torsion-free element*.

Thus m is a torsion free element if and only if $rm = 0$ implies $r = 0$. Notice that in an R-module zero is always a torsion element unless $R = \{0\}$.

6.3. Lemma. *Let R be a commutative ring with 1 and let M be an R-module. Then for $m \in M$ the set*

$$\mathbf{o}(m) = \{r \in R : rm = 0\}$$

is an ideal of R.

Proof. Clearly $0 \in \mathbf{o}(m)$ by Remark 3 on page 71. Let $r_1, r_2 \in \mathbf{o}(m)$ and $r \in R$. We must prove that $r_1 - r_2$ and rr_1 belong to $\mathbf{o}(m)$. Now $(r_1 - r_2)m = r_1 m - r_2 m = 0 - 0 = 0$, and $(rr_1)m = r(r_1 m) = r0 = 0$, as required. (How many of the module axioms have we used?)

6.4. Definition. $\mathbf{o}(m)$ is called the *order ideal* of m.

Remarks. 1. In this terminology an element of M is a torsion element if and only if it has a non-zero order ideal.
2. For general rings we can only say that $\mathbf{o}(m)$ is a *left ideal* (cf. p. 74).

Examples. 1. Consider the cyclic group $\mathbf{Z}_3 = \{[0], [1], [2]\}$. As an Abelian group, $\mathbf{Z}_3$ is a $\mathbf{Z}$-module in the usual way; let us determine $\mathbf{o}([1])$. Since $n[1] = [n]$, we have $n \in \mathbf{o}([1])$ if and only if $3|n$. Hence $\mathbf{o}([1]) = 3\mathbf{Z}$. Thus the element $[1]$, which in the usual group-theoretic sense has order 3, has as order ideal the ideal of $\mathbf{Z}$ generated by 3 (also by -3). The reader will also be able to check that $\mathbf{o}([2]) = 3\mathbf{Z}$.

In general, if A is an arbitrary Abelian group thought of as a Z-module, an element of A is periodic (i.e. has finite order in the group-theoretic sense) if and only if its order ideal is non-zero; in that case its order as a group element coincides with the positive generator of its order ideal. And an element of A has infinite order if and only if its order ideal is the zero ideal. Obviously, the concept 'order ideal' generalizes readily to a module over an arbitrary commutative ring, where the concept 'order of an element' does not.

2. A vector space V over a field **k**, considered as a **k**-module, is torsion-free. For suppose $0 \neq \mu \in \mathbf{k}$ and $\mu v = 0$, where $v \in V$. Then $0 = \mu^{-1} 0 = \mu^{-1}(\mu v) = 1v = v$. Hence 0 is the only torsion vector in V. We shall see later, however, that a finite-dimensional vector space V, made into a **k**$[x]$-module via a linear transformation α, is a torsion module!

3. If R is an integral domain, the module $_R R$ is torsion-free. For $rs = 0$ and $r \neq 0 \Rightarrow s = 0$, and therefore the zero of $_R R$ is its only torsion element.

6.5. Theorem. *Let M be a module over an integral domain R and let T denote the set of torsion elements of M. Then T is a submodule of M, and the quotient module M/T is torsion-free.*

Proof. Clearly $0 \in T$. Let $t_1, t_2 \in T$. Then by definition there exist elements $r_1, r_2 \in R^* = R \smallsetminus \{0\}$ such that $r_i t_i = 0$ for $i = 1, 2$. Hence

$$r_1 r_2 (t_1 - t_2) = (r_2 r_1) t_1 - (r_1 r_2) t_2 = r_2 (r_1 t_1) - r_1 (r_2 t_2)$$
$$= r_2 0 - r_1 0 = 0.$$

Since R has no zero divisors, $r_1 r_2 \in R^*$, and so $t_1 - t_2 \in T$. Finally, if $r \in R$, we have $r_1(rt_1) = r(r_1 t_1) = r0 = 0$, and therefore $rt_1 \in T$. Therefore by 5.3 T is a submodule of M.

To show that M/T is torsion-free let $m + T \in M/T$ and suppose there exists non-zero $r \in R$ such that $r(m + T) = T$. Then $rm \in T$, and so there exists $s \in R^*$ such that $s(rm) = 0$. But $s(rm) = (sr)m$. Since R is an integral domain, $sr \in R^*$, and so $m \in T$. Therefore $m + T = T$, and the zero of M/T is its only torsion element. Hence M/T is torsion-free.

SOME SPECIAL CLASSES OF MODULES

3. Free modules

The concept of a free module over a ring is in many ways a better analogue of a vector space over a field than is that of an arbitrary module. In fact, as we shall see, every module over a field is free – that is why the concept of 'free vector spaces' never arises explicitly in linear algebra. However, although free modules look very much like vector spaces one should guard against the resulting feeling of security, which is by no means always justified.

Free modules will play a big part in analysing the set-up of our main theorems. Every module turns out to be a homomorphic image of a free module, and by using this fact in conjunction with the homomorphism theorems we can answer questions about modules in general by translating them into questions about quotient modules of free modules. These in turn may be studied by examining the submodules which occur as their kernels.

6.6. Definition. Let M be an R-module and let X be a subset of M. We say that X *generates M freely* if

(i) X generates M (as R-module), and

(ii) every map of X into an R-module extends to a R-homomorphism. More explicitly, given any R-module N and map $\phi: X \to N$, there exists an R-homomorphism $\psi: M \to N$ such that $\psi(x) = \phi(x)$ for all $x \in X$.

Any R-module which is freely generated by some subset is called *free*. Any set which freely generates an R-module M is called a *basis* (sometimes free basis) of M.

Remarks. 1. The extending homomorphism ψ is unique. For, if ψ and ψ' are two homomorphisms extending the map ϕ, then the set $\{m \in M : \psi(m) = \psi'(m)\}$ is a submodule of M. Since this set contains X and X generates M, it must be the whole of M. Hence $\psi = \psi'$.

2. Notice that the zero module $\{0\}$ is freely generated by the empty set.

We have chosen this apparently rather abstract definition of freeness because it ties in with the definition of freeness given in other more general contexts. However, to understand free modules better we need a more concrete description of them.

6.7. Definition. A finite, non-empty subset $\{m_1,\ldots,m_t\}$ of an R-module M is said to be *linearly dependent* if there exist elements $r_i \in R$, not all zero, such that $\sum_{i=1}^{t} r_i m_i = 0$. Otherwise the subset is called *linearly independent*. In that case, whenever $\sum_{i=1}^{t} r_i m_i = 0$ we must have $r_1 = \cdots = r_t = 0$. It is convenient to declare the empty set to be linearly independent. For completeness (although we shall not use this) we mention that an infinite set X of elements of M is said to be linearly independent if every finite subset of X is linearly independent.

Thus every subset of a linearly independent set is linearly independent; also, every subset which contains 0 is linearly dependent, unless $R = \{0\}$.

We now put the definition of freeness into perspective.

6.8. Theorem. *Let M be an R-module and let $\{m_1,\ldots,m_s\}$ be a finite subset of M. Then the following statements are equivalent:*

(i) $\{m_1,\ldots,m_s\}$ *generates M freely.*

(ii) $\{m_1,\ldots,m_s\}$ *is linearly independent and generates M.*

(iii) *Every element $m \in M$ is uniquely expressible in the form $m = \sum_{i=1}^{s} r_i m_i$ with $r_i \in R$.*

(iv) *Each m_i is torsion-free, and $M = Rm_1 \oplus \cdots \oplus Rm_s$.*

Proof. (i) $\Rightarrow$ (ii). By definition $\{m_1,\ldots,m_s\}$ generates M. Suppose we have $\sum r_i m_i = 0$ ($r_i \in R$). Let N be the external direct sum ${}_R R \oplus \cdots \oplus {}_R R$ of ${}_R R$ with itself s times, and let $e_i = (0,\ldots,0,1,0,\ldots,0)$ with 1 in the i-th place. Then by definition the map $m_i \to e_i$ extends to a homomorphism ϕ of M into N. We have

$$0 = \phi(0) = \phi(\sum r_i m_i) = \sum r_i \phi(m_i) = \sum r_i e_i = (r_1,\ldots,r_s).$$

Therefore $r_1 = \cdots = r_s = 0$, and this proves that $\{m_1,\ldots,m_s\}$ is linearly independent.

(ii) $\Rightarrow$ (iii). Since $\{m_1,\ldots,m_s\}$ generates M, every element of M can be expressed as an R-linear combination of the m_i. If $\sum r_i m_i = \sum r'_i m_i$ with $r_i, r'_i \in R$, then $\sum (r_i - r'_i) m_i = 0$, and so

SOME SPECIAL CLASSES OF MODULES

$r_i - r'_i = 0$ by definition of linear independence. Hence every element of M is uniquely expressible as a linear combination of the m_i.

(iii) ⇒ (iv). If $r \in R$ and $rm_i = 0$, then, since also $0m_i = 0$, we get $r = 0$ from the uniqueness statement of (iii). Thus each m_i is torsion-free. Clearly $M = \sum Rm_i$. To show that the sum is direct let $m \in Rm_i \cap \sum_{j \neq i} Rm_j$. Then $m = r_i m_i = \sum_{j \neq i} r_j m_j$ for suitable $r_k \in R$, and so by the uniqueness statement we obtain $r_i = 0$. Hence $m = 0$ as required.

(iv) ⇒ (i). It is convenient to go via (iii). Indeed (iv) certainly implies that every element of M can be written in the form $\sum r_i m_i$ with $r_i \in R$. If $\sum r_i m_i = \sum r'_i m_i$, then by 5.15 we obtain $r_i m_i = r'_i m_i$ for each i. Therefore $(r_i - r'_i)m_i = 0$ and since m_i is torsion-free, $r_i - r'_i = 0$. This gives (iii).

Now let N be any R-module and let $m_i \to n_i$ be any map of $\{m_1, \ldots, m_s\}$ into N. If $m \in M$, we have $m = \sum r_i m_i$ where the r_i are uniquely determined elements of R. Define $\phi(m) = \sum r_i n_i$. Notice that this only makes sense because the r_i are uniquely determined by m. Then it is a straightforward matter to check that ϕ is the required homomorphism.

6.9. Corollary. *M is freely generated by s elements if and only if $M \cong {}_R R \oplus \cdots \oplus {}_R R$ with s ${}_R R$'s.*

Proof. It will be convenient to write $({}_R R)^s$ for the external direct sum of ${}_R R$ with itself s times. Let e_i denote the s-tuple whose only non-zero coordinate is a 1 in the s-th place. Then $(r_1, \ldots, r_s) = \sum r_i e_i$, and so $\{e_1, \ldots, e_s\}$ generates $({}_R R)^s$. Since that set is clearly linearly independent, it generates $({}_R R)^s$ freely. Clearly any module isomorphic to a free module is freely generated by the same number of elements.

Conversely, if M is freely generated by $\{m_1, \ldots, m_s\}$, then, since $\{e_1, \ldots, e_s\}$ generate $({}_R R)^s$ freely, there exist homomorphisms $\phi: M \to ({}_R R)^s$ and $\psi: ({}_R R)^s \to M$ sending $m_i \to e_i$ and $e_i \to m_i$ respectively. Then $\phi\psi$ maps e_i to e_i, and so maps every linear combination of the e_i's to itself. Therefore $\phi\psi$ is the identity map of $({}_R R)^s$, and similarly $\psi\phi$ is the identity map on M. Therefore each of ϕ and ψ is an isomorphism, as required.

Remarks. 1. We have only stated the above results for finite generating sets (since this is all we shall need), but it is easy to see that they remain true for arbitrary generating sets.

2. By Corollary 6.9 the module $_R R$ is always freely generated by one element, namely 1.

3. It is a well-known result of linear algebra that, if **k** is a field, then every FG **k**-module (that is vector space with a finite spanning set) has a linearly independent generating set (better known as a basis), and is therefore free. Our definition of a basis for an arbitrary module is therefore consistent with the usual meaning of basis given in the special case of a **k**-module.

4. Warning! Every generating set (spanning set) of a vector space contains a basis for that space. This statement is false for free modules in general. It is not even true for free **Z**-modules. For consider the **Z**-module $_Z\mathbf{Z}$, already seen to be free; it is generated (not freely however) by the set $X = \{2,3\}$, because the submodule generated by X contains $3 - 2 = 1$ which certainly generates $_Z\mathbf{Z}$. However, X is not a basis of $_Z\mathbf{Z}$ (because the equation $3.2 - 2.3 = 0$ shows it is **Z**-dependent), and no proper subset of X is a basis because $\{2\}$, $\{3\}$ and $\emptyset$ generate proper submodules.

Another statement which is true for vector spaces but breaks down for free modules in general is the following: if $\{m_1,\ldots,m_s\}$ is a linearly dependent set then some m_i is a linear combination of the others. The subset X of $_Z\mathbf{Z}$ considered above again provides a counter-example. For neither of the elements 2 and 3 is a **Z**-multiple of the other.

5. Although $\mathbf{Z}_n$ is free as $\mathbf{Z}_n$-module, it is not free as **Z**-module if $n \neq 0$. For every element x of $\mathbf{Z}_n$ satisfies $nx = 0$, and so every non-empty subset of $\mathbf{Z}_n$ is linearly dependent over **Z**. When $\mathbf{Z}_n$ is thought of as $\mathbf{Z}_n$-module, however, the coefficient n becomes zero and the equation $[n]x = 0$ no longer says that x is linearly dependent.

6. One might be tempted (by analogy with vector spaces again) to define the dimension of a free module to be the number of elements in a basis of that module. However, over sufficiently bad rings R there exist free R-modules having bases with different numbers of elements. We shall see in the next chapter that this cannot happen if R is a PID; indeed, if R is any commutative ring with $1 \neq 0$, any

SOME SPECIAL CLASSES OF MODULES

two bases of a free R-module have the same number of elements (cf. Exercise 16 at the end of this chapter).

The following result explains much of the importance of free modules; once again we prove it only in the FG case although it is true in general.

6.10. Theorem. *Every FG R-module is a homomorphic image of a free R-module.*

Proof. Let $M = \sum_{i=1}^{s} Rm_i$ be an R-module generated by a finite set of s elements. Choose a free R-module F with a basis $\{x_1,\ldots,x_s\}$ of s elements. These exist by 6.9; indeed we may take F to be $(_RR)^s$. By definition of freeness the map $x_i \to m_i$ extends to an R-homomorphism $\phi:F \to M$; im ϕ is a submodule of M containing a generating set for M and is therefore M itself. This concludes the proof.

We elaborate a special case of this theorem in the following:

6.11. Theorem. *Let R be a commutative ring with 1 and let $M = Rm$ be a cyclic R-module. Then M is R-isomorphic with the quotient module $_R(R/\mathrm{o}(m))$. Thus two cyclic R-modules are isomorphic if and only if they have the same order ideal.*

Remarks. 1. When we have some object A which can be viewed in various ways and we want to emphasize that we are viewing it as an R-module we do so by writing $_RA$. This is the case with the quotient module $R/\mathrm{o}(m)$ of $_RR$ above. Among other things $R/\mathrm{o}(m)$ is an Abelian group, a ring, an R-module and a module over itself.

2. We have not yet defined the order ideal of a cyclic module. Suppose that $M = Rm$ is a cyclic module over a *commutative* ring R with 1. Then, if $r \in R$ and $rm = 0$, we have $r(sm) = s(rm) = s0 = 0$ for any $s \in R$, and so $rM = \{0\}$. Thus $\mathrm{o}(m) = \{r \in R : rM = \{0\}\}$, and in particular any two generators of M have the same order ideal.

6.12. Definition. *If M is a cyclic R-module over a commutative ring R with 1, the order ideal of any generator of M is called the* order ideal of M.

Proof of Theorem 6.11. The map $r \to rm$ is an R-epimorphism of $_R R$ onto $M = Rm$, as in the proof of 6.10; it is obtained by extending the map $1 \to m$. Its kernel is clearly $\mathbf{o}(m)$. Therefore, by 5.10, $_R(R/\mathbf{o}(m)) \cong M$.

Therefore, if two cyclic R-modules have the same order ideal, they are isomorphic to the same quotient module of $_R R$ and so to each other. The converse is clear.

Remark. Most of this chapter has been conducted under the hypothesis that R is a ring with 1, but occasionally the extra hypothesis of commutativity has been brought in. The reader may find it useful to check where this has been done.

Exercises for Chapter 6

(R will denote a commutative ring with 1 unless otherwise stated.)

1. Let N be a submodule of an R-module M. Show that, if N and M/N are FG, then M is FG.

2. Give an example of an R-module $M = M_1 \oplus M_2$ which is generated by a set X such that $X \cap M_1 = X \cap M_2 = \varnothing$.

3. Find an R-module in which the set of torsion elements is not a submodule. (*Hint*: consider $\mathbf{Z}_n$ for a suitable n.)

4. Prove that submodules and quotient modules of torsion modules are torsion modules. Prove that submodules of torsion-free modules are torsion-free but that quotient modules of torsion-free modules need not be torsion-free.

5. Suppose that the R-module $M = M_1 + M_2$ is the sum of two torsion-free submodules; is M necessarily torsion-free? What is the answer when $M = M_1 \oplus M_2$?

6. Show that $\mathbf{Q}$ is a torsion-free $\mathbf{Z}$-module which is not free.

SOME SPECIAL CLASSES OF MODULES

7. Consider each of the following statements, decide whether it is true or false, and give a proof or a counter-example as appropriate:

   (i) A submodule of a free module is free.
   (ii) A submodule of a free module is torsion-free.
   (iii) A quotient module of a cyclic module is cyclic.
   (iv) A submodule of a cyclic module is cyclic.

8. Let M and N be R-modules freely generated by n elements. Prove that $M \cong N$.

9. Let V be a 3-dimensional vector space over $\mathbf{Z}_2$ considered as a $\mathbf{Z}_2[x]$-module via α, where $\alpha \in \mathrm{End}_{\mathbf{Z}_2} V$ is defined on basis elements as follows:

   $$\alpha(v_1) = v_1 + v_3$$
   $$\alpha(v_2) = v_1 + v_2$$
   $$\alpha(v_3) = v_2 + v_3.$$

   Find $\mathrm{o}(v)$ for each $v \in V$, and deduce that V is a torsion module. Find a non-zero element $f \in \mathbf{Z}_2[x]$ such that $fV = \{0\}$.

10. Let R be a commutative ring with 1, and let J and K be ideals of R. Prove that $_R(R/J) \cong {_R(R/K)}$ if and only if $J = K$.

11. Let M be an R-module freely generated by a set X, and let Y be a subset of X. Show that Y freely generates RY. Show that the direct sum of two free modules is free.

12. Give an example of a $\mathbf{Z}$-module $M = T \oplus F_1 = T \oplus F_2$ where T is the torsion submodule and F_1 and F_2 are distinct non-zero submodules. Prove that in this situation F_1 and F_2 are isomorphic torsion-free modules.

13*. Prove that the module $_RR$ is torsion-free if and only if either $R = \{0\}$ or R is an integral domain. Show that every submodule of $_RR$ is free if and only if either $R = \{0\}$ or R is a PID.

14*. Show that over non-commutative rings, different generators of a cyclic module may have different order left ideals. We outline a possible approach: let $V = \mathbf{k}^2$, where $\mathbf{k}$ is a field, and consider V as a module over the ring $\mathbf{M}_2(\mathbf{k})$ by identifying V with the set of column vectors

$$\begin{bmatrix} x \\ y \end{bmatrix}$$

with $x, y \in \mathbf{k}$, and defining the action of $\mathbf{M}_2(\mathbf{k})$ by matrix multiplication. Show that V is cyclic and that each of

$$\begin{bmatrix} 1 \\ 0 \end{bmatrix} \text{ and } \begin{bmatrix} 0 \\ 1 \end{bmatrix}$$

generate V as an $\mathbf{M}_2(\mathbf{k})$-module. Now calculate the order left ideals of these generators.

15. Let $R \neq \{0\}$ and let M be an R-module freely generated by a set X. Show that no proper subset of X generates M.

16**. Let R be a commutative ring with 1 and let M be a free R-module.

(i) Prove that any two finite bases of M have the same number of elements as follows. Using Chapter 2 Exercise 13 let J be a maximal ideal of R. Show that the R-module M/JM can be viewed as a module for the field R/J (cf. Chapter 5 Exercise 10) and that, if $\{x_1, \ldots, x_s\}$ is a finite R-basis for M, then $\{x_1 + JM, \ldots, x_s + JM\}$ is an R/J-basis of M/JM.

(ii) Show that, if M has a finite basis, then any two bases of M have the same number of elements. By (i) this simply involves showing that M cannot have an infinite basis; do this by showing that some finite subset of such a basis would generate M, or by the method of (i).

(iii) Show that any two infinite bases of an R-module have the same cardinality. Here R may be any ring. A little cardinal arithmetic is required.

PART TWO

Direct decomposition of a finitely-generated module over a principal ideal domain

Except where otherwise stated, all rings appearing in this section will be assumed to be principal ideal domains

CHAPTER SEVEN

Submodules of free modules

1. The programme

Our aim in this section is to prove a decomposition theorem. The ingredients of the theorem are

(a) a principal ideal domain R, and
(b) a finitely-generated module M over R.

The conclusions of the theorem are:

M can be expressed as an internal direct sum

$$M = M_1 \oplus M_2 \oplus \cdots \oplus M_t$$

such that

(i) each $M_i = Rm_i$ is a *cyclic* submodule, and
(ii) $o(m_1) \supseteq o(m_2) \supseteq \cdots \supseteq o(m_t)$.

Our method is to consider an epimorphism $\theta : F \to M$ from a free module F onto M. We know that $\ker \theta \, (= N$ say) is a submodule of F and by the first isomorphism theorem for modules (5.10) that M is isomorphic to the quotient module F/N. So we can get at the structure of M by studying F/N. It turns out that submodules of finitely-generated free modules are free – see 7.8 below – and therefore, in particular, that N is free. We then show by the following result how it is possible to choose a basis for F coupled in a very special way with a basis for N.

7.1. Theorem. *Let R be a PID, F a free R-module of finite rank s, and N a submodule of F. Then there exists a basis $\{f_1, \ldots, f_s\}$ of F and elements $d_1, \ldots, d_s \in R$ such that*

(a) *the non-zero elements of* $\{d_1 f_1, \ldots, d_s f_s\}$ *form a basis for* N, *and*

(b) $d_1 | d_2 | \cdots | d_s$.

Remarks. 1. This is the best analogue we have in the present context of the well-known fact that, if U is a subspace of a finite-dimensional vector space V over some field, then any basis $\{f_1, \ldots, f_t\}$ of U can be extended to a basis $\{f_1, \ldots, f_t, f_{t+1}, \ldots, f_s\}$ of V. In that case we have $d_1 = \cdots = d_t = 1, d_{t+1} = \cdots = d_s = 0$. Our theorem above says that *certain* bases of N arise in a nice way from bases of F. But the fact that not all bases of N arise in this way is an indication that the general situation is not as straightforward as for vector spaces.

2. Condition (b) translated into the language of ideals says that $d_1 R \supseteq d_2 R \supseteq \cdots \supseteq d_s R$. If $d_i = 0$ for any i, we have $d_j = 0$ for all $j = i, i+1, \ldots, s$.

Theorem 7.1 will be our main concern in this chapter. We shall prove it by converting it into a problem about matrices over R. In Chapter 8 we shall use 7.1 to prove the decomposition theorem outlined above. Then we shall examine the question of uniqueness and look for further refinements. Because the decomposition theorem is the central result of this book, we feel justified in offering an alternative treatment of it. Chapter 9 is devoted to this. The proof we give there is more direct, and therefore shorter, but it is conceptually more demanding and in some ways less informative.

2. Free modules – bases, endomorphisms and matrices

The reader is no doubt familiar with the well-known correspondence between endomorphisms of a finite-dimensional vector space over a field **k** (in other words, linear transformations $V \to V$) and $n \times n$ matrices over **k**. The description of this correspondence extends easily to the case of a finitely-generated free module over a PID. We shall indicate how this goes, but in view of the familiarity of the arguments we shall be fairly brief.

In the statement of 7.1 we used the word 'rank' to describe the number of elements in a basis of a free module. But before we can give the formal definition we need to know that it is a genuine

SUBMODULES OF FREE MODULES

invariant; in other words, that any two bases of a free module have the same number of elements.

7.2. Theorem. *Let F be a module over a PID R, and suppose that F is freely generated by a finite set of n elements. Then every basis of F contains exactly n elements.*

Proof. We prove first, by induction on n, that no finite linearly independent subset of F can contain more than n elements.

This is certainly true if $n = 0$, since then $F = \{0\}$, and the only linearly independent subset of F is the empty set. In the case $n = 1$, we have $F = Rx$. Let rx, sx be elements of F. Then the relation $s(rx) - r(sx) = 0$ tells us that $\{rx, sx\}$ is linearly dependent, except in the case $r = s = 0$; but then $\{rx, sx\} = \{0\}$, and so is certainly linearly dependent. Therefore no linearly independent subset of F can contain two elements. Since every subset of a linearly independent set is linearly independent, no linearly independent subset of F can contain more than two elements either.

Now suppose that $n > 1$ and that $F = Rf_1 \oplus \cdots \oplus Rf_n$. Let $\bar{F} = Rf_2 \oplus \cdots \oplus Rf_n$, and let $X = \{x_1, \ldots, x_m\}$ be a subset of F with $m > n$. If $X \subseteq \bar{F}$, then X is certainly linearly dependent by induction. Otherwise we may suppose without loss of generality that $x_1 \notin \bar{F}$. Now

$$F/\bar{F} \cong Rf_1 \qquad (1)$$

which is freely generated by one element, and it follows that for $i \geqslant 2$ the set $\{x_1 + \bar{F}, x_i + \bar{F}\}$ is linearly dependent. Therefore there exist elements $r_i, s_i \in R$ and not both zero such that $r_i(x_1 + \bar{F}) + s_i(x_i + \bar{F}) = \bar{F}$, that is, $r_i x_1 + s_i x_i \in \bar{F}$. Now, if $s_i = 0$, we obtain $r_i(x_1 + \bar{F}) = \bar{F}$. But $x_1 + \bar{F}$ is a non-trivial element of the module $F/\bar{F}$ which is isomorphic to $_R R$ by (1); by Example 3 on p. 88 $_R R$ is torsion-free, and therefore r_i is also zero, a contradiction. Hence $s_i \neq 0$ for $i = 2, \ldots, m$. Set $y_i = r_i x_1 + s_i x_i$; then we have found $m - 1$ elements $y_2, \ldots, y_m$ of $\bar{F} = Rf_2 \oplus \cdots \oplus Rf_n$ which is freely generated by the $n - 1$ elements $\{f_2, \ldots, f_n\}$. Since $m - 1 > n - 1$ our induction shows that these elements are linearly dependent. Therefore there exist elements $t_2, \ldots, t_m \in R$, not all zero, such that $\sum_{i=2}^{m} t_i y_i = 0$. If we express the left-hand side as a linear combination of $x_1, \ldots, x_n$, the coefficient of x_i for $i \geqslant 2$ is $t_i s_i$. Since at least one t_i is non-zero and no s_i is zero, it follows that at least one of these coefficients is non-zero. Hence $\{x_1,\ldots,x_m\}$ is linearly dependent. This proves the assertion with which we began the proof.

It follows that if $\{u_1,\ldots,u_k\}$ is another finite basis of F, then $n \geqslant k$. However, by symmetry $k \geqslant n$, and so we have equality. It remains remotely possible that F might have an infinite basis, Z say. In that case, let $z_1,\ldots, z_t$ be finitely many elements of Z, and let $F^* = \sum_{i=1}^{t} Rz_i$. Given any map $z_i \to m_i$ of $\{z_1,\ldots,z_t\}$ into an R-module M, we can extend it to a homomorphism of F^* into M as follows: first extend it to Z by demanding that the remaining elements of Z be mapped to zero, and then extend to a homomorphism of the whole of F into M using the fact that Z freely generates F (recall Definition 6.6). It follows that the set $\{z_1,\ldots,z_t\}$ freely generates F^* and is therefore linearly independent by 6.8. Hence, by the statement at the beginning of the proof we have $t \leqslant n$. But since Z is infinite, we can certainly arrange that $t > n$ and so obtain a contradiction. This concludes the proof.

In view of this result, we can now make the following definition.

7.3. Definition. Let F be a free module (over a PID) with a finite basis. Then the number of elements in a basis of F is called the *rank* of F.

Remarks. 1. For vector spaces the rank is obviously the same as the dimension in the usual sense.
2. When we speak of a basis of F in the sequel, we shall often mean an ordered basis, by which we understand simply a set of elements forming a basis and written down in some specified order. Two bases which consist of the same elements written down in a different order are then considered to be different. Rather than introduce some notation to distinguish between ordered and unordered bases, we leave the reader to deduce from the context which we mean; it should be noted, however, that in dealing with matrices of endomorphisms as below the ordering of the basis is always important.

SUBMODULES OF FREE MODULES

Now let F be a free R-module of finite rank $s > 0$ (where R, as we have said, henceforth denotes a PID), and let $\mathbf{f} = \{f_1, \ldots, f_s\}$ be a basis of F. Then, if $\alpha \in \text{End}_R F$, by 6.8 we have

$$\alpha(f_i) = \sum_{j=1}^{s} a_{ji} f_j \quad (i = 1, 2, \ldots, s) \tag{2}$$

for certain uniquely determined elements $a_{ji} \in R$. Thus α determines uniquely an $s \times s$ matrix $A = (a_{kl})$ with entries in R; the i-th column of A consists of the coefficients of $\alpha(f_i)$ with respect to the basis $\mathbf{f}$. Conversely, given an arbitrary $s \times s$ matrix $A = (a_{kl})$ with entries in R, the definition of freeness tells us that there is a unique endomorphism of F whose effect on $f_1, \ldots, f_s$ is given by (2). So the correspondence between endomorphisms and matrices expressed by (2) is bijective.

We can make $\text{End}_R F$ into a ring in the usual way, that is, by defining the sum and product of each pair $\alpha, \beta \in \text{End}_R F$ as follows:

$$(\alpha + \beta)(x) = \alpha(x) + \beta(x); \quad (\alpha\beta)(x) = \alpha(\beta(x)) \quad \text{for all } x \in F.$$

It is easy to verify that $\alpha + \beta$ and $\alpha\beta$ are endomorphisms of F, and that these operations turn $\text{End}_R F$ into a ring. If β corresponds to the matrix (b_{kl}), we have $\beta(f_i) = \sum_j b_{ji} f_j$. Therefore

$$(\alpha + \beta)(f_i) = \alpha(f_i) + \beta(f_i) = \sum_j a_{ji} f_j + \sum_j b_{ji} f_j = \sum_j (a_{ji} + b_{ji}) f_j,$$

and

$$(\alpha\beta)(f_i) = \alpha(\beta(f_i)) = \alpha\left(\sum_j b_{ji} f_j\right) = \sum_j b_{ji} \alpha(f_j)$$

$$= \sum_j b_{ji} \left(\sum_k a_{kj} f_k\right) = \sum_k \left(\sum_j a_{kj} b_{ji}\right) f_k.$$

Thus the matrix corresponding to $\alpha + \beta$ is $(a_{kl} + b_{kl})$ and that corresponding to $\alpha\beta$ has (k, l)-entry $\sum_j a_{kj} b_{jl}$. Therefore, if the sum and product of matrices are defined in the usual way, the map which associates with each endomorphism its matrix is a ring endomorphism of $\text{End}_R F$ onto $\mathbf{M}_s(R)$. Indeed, this is precisely why matrix addition and multiplication are defined as they are. Notice that this isomorphism is constructed with respect to some particular selected basis of F; different bases will in general correspond to different isomorphisms.

Now an endomorphism α of F is an automorphism if and only if there exists another endomorphism β such that

$$\alpha\beta = \beta\alpha = 1_F, \qquad (3)$$

the identity map on F. The matrix of 1_F is evidently the usual $s \times s$ identity matrix. Because of our isomorphism between $\mathrm{End}_R F$ and $M_s(R)$, (3) above is equivalent to

$$AB = BA = 1_s, \qquad (4)$$

where A and B are the matrices of α, β respectively with respect to **f** and 1_s is the $s \times s$ identity matrix.

7.4. Definition. An $s \times s$ matrix A over R is called *invertible* if there exists an $s \times s$ matrix B over R satisfying (4). (Invertible matrices are often called *non-singular*, a terminology particularly common in the case when R is a field.) The equivalence of (3) and (4) tells us that automorphisms of F correspond precisely to invertible matrices in the isomorphism between $\mathrm{End}_R F$ and $M_s(R)$. The invertible matrices are in an obvious sense the units of $M_s(R)$.

As with vector spaces, in dealing with free modules we often want to pass from one basis to another and the above considerations tell us how to do so. We already know from 7.2 that any basis of F has s elements. Let $\mathbf{f}^* = \{f_1^*, \ldots, f_s^*\}$ be a set of s elements, and consider the question: under what circumstances is $\mathbf{f}^*$ a basis of F? We have

$$f_i^* = \sum_{j=1}^{s} a_{ji} f_j \quad (i = 1, \ldots, s),$$

where $A = (a_{kl})$ is a certain $s \times s$ matrix over R. We know from the definition of freeness that there is a unique R-endomorphism α of F defined by $\alpha(f_i) = f_i^*$ for $i = 1, \ldots, s$. From the above equation we see that the matrix of α with respect to **f** is A. With this notation we then have

7.5. Lemma. *The following statements are equivalent:*
  (i) $\mathbf{f}^*$ *is a basis of* F.
  (ii) α *is an automorphism of* F.
  (iii) A *is an invertible matrix.*

SUBMODULES OF FREE MODULES

Proof. Since we have already shown the equivalence of the last two statements, we have only to prove that (i) is equivalent to (ii).

It is clear that if α is an automorphism of F then $\mathbf{f}^*$ will be a basis. For, if $m_1,\ldots, m_s$ are given elements of some R-module M, then by definition of freeness there exists an R-homomorphism $\theta: F \to M$ such that $\theta(f_i) = m_i$ for $1 \leq i \leq s$. But then $\theta\alpha^{-1}$ is an R-homomorphism $F \to M$ mapping each f_i^* to m_i.

Conversely, if $\mathbf{f}^*$ is a basis, then there exists an R-endomorphism β of F such that $\beta(f_i^*) = f_i$ $(1 \leq i \leq s)$. Clearly $\beta\alpha = \alpha\beta = 1_F$, and so α is an automorphism.

Thus changes of basis of free modules over a PID are effected by invertible matrices just as in the case of vector spaces.

It cannot be emphasized too much that the above discussion of the representation of endomorphisms of F by matrices is all conducted with respect to some fixed basis $\mathbf{f}$ of F. However, it is often important to know what happens if we pass to some new basis $\mathbf{f}^*$ of F – how the matrix of an endomorphism α with respect to $\mathbf{f}^*$ is related to its matrix with respect to $\mathbf{f}$. This question is not hard to answer. For let ξ be the automorphism of F sending f_i to f_i^*. Then the matrix (a_{kl}^*) of α with respect to $\mathbf{f}^*$ is given by

$$\alpha(f_i^*) = \sum_j a_{ji}^* f_j^*,$$

or

$$\alpha\xi(f_i) = \sum_j a_{ji}^* \xi(f_j).$$

Therefore

$$\xi^{-1}\alpha\xi(f_i) = \sum_j a_{ji}^* f_j$$

which means that $A^* = (a_{kl}^*)$ is the matrix of $\xi^{-1}\alpha\xi$ with respect to $\mathbf{f}$. Thus

$$A^* = X^{-1} A X$$

where X is the matrix of ξ with respect to $\mathbf{f}$; as we have seen above, this is the matrix which expresses the f_j^* in terms of the f_i.

We conclude this section with a remark on determinants. The *determinant* $\det X$ of a square matrix X over a commutative ring with 1 can be defined exactly as for a square matrix over a field, and the usual elementary properties of determinants carry over. The reader can easily check this by consulting the development of determinants given in any textbook on linear algebra and examining the proofs to be found there. For the moment we need only the following two facts:

(i) If $X, Y \in \mathbf{M}_s(R)$, then $\det XY = \det X . \det Y$.

(ii) If $X \in \mathbf{M}_s(R)$, then $X . \operatorname{adj} X = \operatorname{adj} X . X = \det X . 1_s$, where $(\operatorname{adj} X)_{ij} = X_{ji}$, the cofactor of x_{ji} in X.

It is easy to deduce from these facts the following characterization of invertible matrices over R:

7.6. Lemma. *Let R be a commutative ring with 1, and let $X \in \mathbf{M}_s(R)$. Then X is invertible if and only if $\det X$ is a unit in R.*

Proof. First suppose that X is invertible. Then $XY = 1_s$ for some $Y \in \mathbf{M}_s(R)$. Taking determinants and using (i) gives $(\det X).(\det Y) = \det 1_s = 1$. Hence $\det X$ is a unit in R.

Conversely, suppose that $\det X$ is a unit in R. Then in (ii) we may divide by $\det X$ to conclude that $XY = YX = 1_s$, where $Y = (\det X)^{-1} . \operatorname{adj} X$. Hence X is invertible.

Thus, for example, the invertible elements of $\mathbf{M}_s(\mathbf{Z})$ are those of determinant ± 1; the invertible elements of $\mathbf{M}_s(\mathbf{k}[x])$ are those whose determinant belongs to $\mathbf{k}^*$.

3. A matrix formulation of Theorem 7.1

In this section we reformulate 7.1 in the language of matrices. But before doing so we need to give the promised proof that submodules of free modules of finite rank (over a PID) are free. We make use of the following lemma which expresses a property of free modules which is very important in more advanced contexts and which we shall require several times in the sequel. This is the 'splitting property', so called because it says that under certain conditions a module splits into a direct sum of two submodules.

7.7. Lemma. *Let M be an R-module, let F be a free R-module of finite rank, and let $\phi : M \to F$ be an epimorphism. Then M has a submodule $F^* \cong F$ such that $M = F^* \oplus \ker \phi$.*

Remark. Although we have only stated this result for free modules of finite rank over a PID, it is true for arbitrary free modules by

SUBMODULES OF FREE MODULES

the same argument. We have stated it under the more restricted hypotheses in the interests of simplicity, and the reader who prefers to do so may ignore the generalization.

Proof. Let $\{f_1, \ldots, f_s\}$ be a basis of F. Since ϕ is an epimorphism, there exist elements $m_i \in M$ such that $\phi(m_i) = f_i$ for $1 \leq i \leq s$. By definition of freeness there is an R-homomorphism $\psi: F \to M$ such that $\psi(f_i) = m_i$ for $1 \leq i \leq s$. Then $\phi\psi(f_i) = f_i$ for all i, and so

$$\phi\psi = 1_F. \tag{5}$$

Let $F^* = \psi(F)$. We claim that this is the required submodule. By (5) we have $\phi(F^*) = \phi\psi(F) = F$. Hence, if $m \in M$, then $\phi(m) = \phi(m^*)$ for some $m^* \in F^*$. Therefore $\phi(m - m^*) = 0$ and $m - m^* \in K = \ker \phi$. Thus $m \in K + F^*$, and it follows that $K + F^* = M$. Now let $n \in K \cap F^*$. Then $n = \psi(f)$ for some $f \in F$. Since $n \in K$, we have $0 = \phi(n) = \phi\psi(f) = f$ from (5). Hence $n = 0$, $K \cap F^* = \{0\}$, and therefore $M = K \oplus F^*$. Finally, since $\phi(F^*) = F$ and the kernel of ϕ restricted to F^* is $F^* \cap K = \{0\}$, it follows that ϕ maps F^* isomorphically onto F.

7.8. Theorem. *If R is a PID and F is a free R-module of finite rank s, then every submodule of F is free of rank $\leq s$.*

Proof. The proof is conducted by induction on the rank s of F. If $s = 0$, then $F = \{0\}$ is freely generated by the empty set and F is the only submodule of itself in this case. If $s = 1$, then $F \cong {}_R R$ by 6.9 and the submodules of F correspond to ideals J of R. Since R is a PID, we have $J = Ra$ for some $a \in J$. If $a = 0$, then $J = \{0\}$ which is a free R-module of rank 0. If $a \neq 0$, then the map $r \to ra$ is an R-module isomorphism of ${}_R R$ onto $Ra = J$. Hence $J \cong {}_R R$ which is free of rank 1.

Now assume that $s > 1$ and that the theorem holds for free modules of rank $< s$. Let $\mathfrak{f} = \{f_1, \ldots, f_s\}$ be a basis of F. Then $F = Rf_1 \oplus \cdots \oplus Rf_s$ by 6.8. Let N be a submodule of F and set $\bar{F} = Rf_2 \oplus \cdots \oplus Rf_s$, a free submodule of rank $s - 1$. Then by induction $\bar{F} \cap N$ is free of rank $\leq s - 1$.

Now $F/\bar{F} = Rf_1 \oplus \bar{F}/\bar{F} \cong Rf_1/Rf_1 \cap \bar{F} = Rf_1/\{0\} \cong Rf_1$ by the isomorphism theorem 5.11, hence $F/\bar{F}$ is free of rank 1. Let ν be the natural homomorphism of F onto $F/\bar{F}$. Then the restriction $\bar{\nu}$ of ν to N maps N onto some submodule of $F/\bar{F}$, and this submodule will be free of rank 1 or 0 by the case $s = 1$. Since $\ker \bar{\nu} = \bar{F} \cap N$, we obtain from Lemma 7.7 that

$$N = L \oplus (\bar{F} \cap N),$$

where L is free of rank 0 or 1. If $L = \{0\}$, then $N = \bar{F} \cap N$ is free of rank $\leq s - 1$. If $L = Rx$ is free of rank 1, then $N = Rx \oplus Rg_1 \oplus \cdots \oplus Rg_t$, where $\{g_1, \ldots, g_t\}$ is a basis of $\bar{F} \cap N$. Since $t \leq s - 1$, we find that N is free of rank $\leq s$, as required.

Let us now return to the situation of 7.1 where N is a submodule of a free module F of finite rank, s say, and assume for the moment that F and N are non-zero. Let $\mathbf{f} = \{f_1, \ldots, f_s\}$ be a basis of F, and let $\mathbf{n} = \{n_1, \ldots, n_t\}$ be a basis of N – such a basis exists by 7.8. We have

$$n_i = \sum_{j=1}^{s} a_{ji} f_j \quad (i = 1, 2, \ldots, t),$$

since the n_i's belong to F. Here the a_{kl} are certain uniquely determined elements of R. $A = (a_{kl})$ is then a certain $s \times t$ matrix uniquely determined by specifying the (ordered) bases $\mathbf{f}$ and $\mathbf{n}$ of F and N respectively. It will be called the *matrix of* $\mathbf{n}$ *with respect to* $\mathbf{f}$. We now consider what happens if we take new bases $\mathbf{f}^*$ and $\mathbf{n}^*$ for F and N. How is the matrix of $\mathbf{n}^*$ with respect to $\mathbf{f}^*$ related to A?

We know from §2 of this chapter that the new bases are given by certain invertible matrices, viz.

$$f_i^* = \sum_{j=1}^{s} x_{ji} f_j,$$

and

$$n_i^* = \sum_{j=1}^{t} y_{ji} n_j,$$

where $X = (x_{kl})$ and $Y = (y_{kl})$ are invertible matrices over R of size $s \times s$ and $t \times t$ respectively. The f_i are expressed in terms of the f_j^* by means of the matrix X^{-1}, since, if $X^{-1} = (\hat{x}_{kl})$, we have

$$\sum \hat{x}_{ji} f_j^* = \sum \hat{x}_{ji} x_{kj} f_k = \sum x_{kj} \hat{x}_{ji} f_k = \sum \delta_{kl} f_k = f_i.$$

Here δ_{ki} is the Kronecker delta, and we are using the convention that summation is over all suffices appearing twice.

SUBMODULES OF FREE MODULES

Therefore we have

$$n_i^* = \sum y_{ji} n_j = \sum y_{ji} a_{kj} f_k = \sum y_{ji} a_{kj} \hat{x}_{lk} f_l^*$$
$$= \sum \hat{x}_{lk} a_{kj} y_{ji} f_l^*$$
$$= \sum (X^{-1} A Y)_{li} f_l^*.$$

Thus the matrix of $\mathbf{n}^*$ with respect to $\mathbf{f}^*$ is

$$A^* = X^{-1} A Y.$$

Therefore, by suitable changes of basis in F and N, we can replace A by any matrix related to it as above, where X and Y may be arbitrary invertible matrices over R of appropriate size. This prompts the following definition.

7.9. Definition. Let A and B be two matrices over R of the same size. Then B is said to be *equivalent* to A (over R) if there exist invertible matrices X and Y over R (of the appropriate size) such that
$$B = XAY.$$

It is easy to verify that this relation of 'equivalence' is in fact an equivalence relation.

We shall now show how the foregoing discussion enables us to reduce 7.1 to the following theorem about matrices.

7.10. Theorem. *Any $s \times t$ matrix A with entries in a PID R is equivalent over R to a matrix $\mathrm{diag}(d_1,\ldots,d_u)$ such that $d_1 | \cdots | d_u$.*

Here $\mathrm{diag}(d_1,\ldots,d_u)$ denotes the $s \times t$ matrix having the elements $d_1, d_2, \ldots$ on its diagonal, that is in the $(1,1), (2,2), \ldots, (u,u)$ places ($u = \min\{s, t\}$), and zeros elsewhere.

It is a simple matter to deduce 7.1 from 7.10. For let N and F be as in 7.1. If $N = \{0\}$, we take any basis of F and all the d_i to be zero. Otherwise, let $\mathbf{n}$ and $\mathbf{f}$ be bases of N and F as above, and let A be the matrix of $\mathbf{n}$ with respect to $\mathbf{f}$. Then by 7.10 there exist invertible matrices X^{-1} and Y over R such that

$$X^{-1} A Y = \mathrm{diag}(d_1,\ldots,d_u),$$

where $d_1 | \cdots | d_u$. X and Y determine new bases $\mathbf{f}^*$ and $\mathbf{n}^*$ of F and N as above, and the matrix of $\mathbf{n}^*$ with respect to $\mathbf{f}^*$ is $\mathrm{diag}(d_1,\ldots,d_u)$. Thus

$$n_1^* = d_1 f_1^*, \ldots, n_u^* = d_u f_u^*$$

is a basis for N. If we now define $d_{u+1} = \cdots = d_s = 0$, remembering that $u \leq s -$ (in fact, u will be t in this situation), we obtain precisely the conclusion of 7.1.

Therefore our aim will now be to prove 7.10. Consequently, we can forget about the modules F and N for a while and concentrate on matrices.

4. Elementary row and column operations

What follows in this paragraph will probably be quite familiar in the more restricted setting of matrices with entries in a field. First, we define a list of special square matrices with entries in R (there is no need to specify their size):

(i) F_{ij} is the matrix obtained from the identity matrix by interchanging row i and row j.

(ii) $G_i(u)$ is the *diagonal* matrix with a unit u of R in the i-th row and 1's elsewhere on the diagonal.

(iii) $H_{ij}(r)$, for any $r \in R$ and $j \neq i$, is the matrix obtained from the identity matrix by adding row j multiplied by r to row i. Thus $H_{ij}(r)$ has 1's on the diagonal, r in the (i, j) place, and zeros elsewhere.

(iv) $\bar{H}_{ij}(r)$ is defined in the same way as $H_{ij}(r)$ with the word 'row' replaced by 'column'. In fact $\bar{H}_{ij}(r) = H_{ji}(r)$, but it is useful to have both notations.

We have $\det F_{ij} = -1$, $\det G_i(u) = u$, $\det H_{ij}(r) = \det \bar{H}_{ij}(r) = 1$. Thus the determinants of these matrices are units of R, and therefore by 7.6 the matrices are all invertible.

7.11. Lemma. *The effect of premultiplying a given matrix of the appropriate size*

   (1) *by F_{ij} is to interchange row i and row j,*
   (2) *by $G_i(u)$ is to multiply row i by u, and*
   (3) *by $H_{ij}(r)$ is to add r times row j to row i.*

The effect of postmultiplying a given matrix of the appropriate size

   (4) *by F_{ij} is to interchange column i and column j,*
   (5) *by $G_i(u)$ is to multiply column i by u, and*
   (6) *by $\bar{H}_{ij}(r)$ is to add r times column j to column i.*

SUBMODULES OF FREE MODULES

Proof. Rather than set up some elaborate notation to prove these simple facts, we prefer to leave the reader to convince himself (on a piece of scrap paper) that they are true.

7.12. Definition. The activities described in 7.11, (1)–(3), are known as the *elementary row operations* on a matrix, and those of 7.11, (4)–(6), as the *elementary column operations*.

It is a useful rule of thumb that to carry out an elementary row operation one has only to perform that operation on the appropriate identity matrix and premultiply by the result; and similarly for column operations with postmultiplication. Since the matrices performing the tasks are invertible, any of these elementary operations will transform a matrix into an equivalent one. Therefore we can prove that two matrices are equivalent by showing that one can be transformed into the other by a sequence of elementary operations, letting the actual matrix operators recede into the background. If we do need to keep track of the transforming matrices for any reason, we just have to perform the appropriate sequence of row operations on an identity matrix to get the pre-operator, and the column operations on another identity matrix to obtain the post-operator. Indeed, the elementary row operations are performed by premultiplying successively by elementary matrices $X_1, \ldots, X_k$; their cumulative effect is produced by premultiplying by $X = X_k \ldots X_1$, and this is just the matrix obtained from the identity by the same sequence of row operations.

5. Proof of 7.10 for Euclidean domains

It is instructive to prove 7.10 first for the special case of a Euclidean domain, since it is this situation with which we shall be concerned in the applications. In addition, the argument for ED's is somewhat simpler than that for arbitrary PID's.

We shall therefore take an arbitrary $s \times t$ matrix A over an ED R (equipped with a Euclidean function ϕ), and shall show how to reduce A by means of elementary row and column operations to a matrix $\mathrm{diag}(d_1, \ldots, d_u)$ where $u = \min\{s, t\}$ and $d_1 | d_2 | \cdots | d_u$. This will prove 7.10 in the special case when R is an ED.

The first stage of the reduction. Our aim here will be to reduce A to an equivalent $s \times t$ matrix C of the special form

$$C = \begin{bmatrix} d_1 & 0 \ldots 0 \\ \hline 0 & \\ \vdots & C^* \\ 0 & \end{bmatrix} \quad \text{where } d_1 \text{ divides each entry of } C^*. \qquad (\pounds)$$

We shall describe a finite sequence of elementary row and column operations which, when performed on A, either yields a matrix of the form ($\pounds$) or else leads to an $s \times t$ matrix $B = (b_{ij})$ satisfying the condition

$$\phi(b_{11}) < \phi(a_{11}). \qquad (\$)$$

In the latter case we go back to the beginning and apply the sequence of operations again. Either we reach ($\pounds$), in which case we stop, or we reach ($\$$) again, in which case the ϕ-value of the leading entry is reduced still further and we continue. It is clear that we must reach ($\pounds$) after a finite number of steps, otherwise each application of our sequence of operations leads back to ($\$$) and the ϕ-values of the leading entries of the matrices obtained form a strictly decreasing infinite sequence of non-negative integers. But of course there can be no such sequence.

The sequence of operations is as follows. If A is the zero matrix, we are already at ($\pounds$); otherwise A has a non-zero entry, and by suitable interchanges of rows and columns this can be moved to the leading position. We therefore assume $a_{11} \neq 0$ and consider the following three possibilities:

Case 1. There is an entry a_{1j} in the first row such that $a_{11} \nmid a_{1j}$. By the properties of Euclidean domains we can write

$$a_{1j} = a_{11} q + r,$$

where either $r = 0$ or $\phi(r) < \phi(a_{11})$. Since $a_{11} \nmid a_{1j}$, we must have $r \neq 0$, and so $\phi(r) < \phi(a_{11})$. Subtract q times the first column from the j-th column and then interchange the first and j-th columns. This replaces the leading entry a_{11} by r and so achieves ($\$$).

Case 2. There is an entry a_{i1} in the first column such that $a_{11} \nmid a_{i1}$. In this case proceed as in case 1, operating with rows instead of columns, to reach ($\$$).

Case 3. a_{11} divides every entry in the first row and first column. In this case, by subtracting suitable multiples of the first column from the other columns, we can replace all the entries in the first row, other than a_{11} itself, by zeros. Similarly, subtracting multiples of the first row from the others, we get a matrix of the form

$$D = \begin{bmatrix} a_{11} & 0\ldots 0 \\ \hline 0 & \\ \vdots & D^* \\ 0 & \end{bmatrix}.$$

If a_{11} divides every entry of D^*, we have reached (£) and all is well. If not, there is an entry, say d_{ij}, such that $a_{11} \nmid d_{ij}$. In that case we add the i-th row to the top row; this brings us back to case 1 and this in turn then leads us to ($).

We have thus arranged that the outcome in each of the three cases is a matrix equivalent to A which either has the form (£) or satisfies ($). Repeated application now brings us to the promised (£) after a finite number of steps, thereby completing the first stage of the reduction.

The conclusion of the reduction. It is easy to see how to proceed, for on reaching (£) we have effectively reduced the size of the matrix with which we are dealing. We can then apply our process to the submatrix C^*, reducing its size still further, and so on, leaving a trail of diagonal elements as we go. There are only two points which need to be made. The first is that any elementary operation on C^* corresponds to an elementary operation on C which does not affect the first row and column. The second is that any elementary operation on C^* gives a new matrix whose entries are linear combinations of the old ones; these new entries will therefore still be divisible by d_1. Therefore, in due course, we reach a matrix $\mathrm{diag}(d_1,\ldots,d_u)$ with $d_1|d_2|\cdots|d_u$, as claimed. To illustrate this process we shall work out the full details of a numerical example at the end of the chapter.

Remark. The matrices $G_i(u)$ mentioned in §4 were only included there for completeness. The operations to which they correspond are usually included in the list of elementary operations but we have not needed them above. They are often important in particular cases – for example, if R is a field, the use of these operations allows us to replace all the non-zero elements d_i by 1, and if $R = \mathbf{Z}$, we can use these operations to make all the non-zero d_i's positive.

6. The general case

The pattern of argument here is not very different from that of §5. The main difference is that elementary row and column operations are no longer sufficient to effect the reduction and another kind of 'secondary operation' has to be brought into play.

If we wish to mimic §5, the first task is to find something to play the part of the Euclidean function. We do this by introducing a 'length function' λ on R^*, where R is our PID. If $r \in R^*$, then we know from 4.14 that r can be written in the form

$$r = up_1 \ldots p_n,$$

where u is a unit, the p_i are primes in R, and $n \geq 0$. Certain features of this expression are unique, among them the integer n. We define $\lambda(r) = n$ and call this the *length* of r. Clearly

$$\lambda(rr') = \lambda(r) + \lambda(r') \quad \text{if } r, r' \in R^*. \tag{6}$$

We now show how to reduce an arbitrary $s \times t$ matrix A over R to a diagonal matrix of the required type by a sequence of operations each of which corresponds to multiplication by an invertible matrix.

The first stage of the reduction. As before, this consists of successive applications of a certain sequence of operations, so chosen that each application either leads to (£) or to the condition obtained by replacing ϕ by the length function λ in ($).

Modification to the sequence of operations is only needed in cases 1 and 2, and it will suffice to explain what happens in case 1. Here we have $a_{11} \neq 0$ and $a_{11} \nmid a_{1j}$ for some j with $1 < j \leq t$. By interchanging columns we may suppose $j = 2$; this is just a notational convenience. Let d be an hcf of a_{11} and a_{12} – such a thing exists by 4.19. Then we have

$$a_{11} = dy_1, \qquad a_{12} = dy_2, \tag{7}$$

SUBMODULES OF FREE MODULES

where, since $a_{11} \nmid a_{12}$, y_1 is not a unit. Therefore $\lambda(y_1) \geqslant 1$, and so from (6) we have

$$\lambda(d) < \lambda(a_{11}). \tag{8}$$

Since $Ra_{11} + Ra_{12} = Rd$ (by 4.19), we can write

$$d = x_1 a_{11} + x_2 a_{12}$$

for some $x_1, x_2 \in R$. Then from (7) we have

$$d = d(x_1 y_1 + x_2 y_2),$$

whence

$$x_1 y_1 + x_2 y_2 = 1.$$

Therefore the determinant of the $t \times t$ matrix

$$S = \begin{bmatrix} x_1 & -y_2 & 0 \\ x_2 & y_1 & \\ \hline 0 & & 1_{(t-2)} \end{bmatrix}$$

is 1, and this matrix is invertible by 7.6.

Consider now the matrix AS. It is equivalent to A and its leading entry is $x_1 a_{11} + x_2 a_{12} = d$. Therefore by (8) it is a matrix satisfying ($), or rather what we get from ($) by replacing the function ϕ by λ.

The conclusion of the reduction. This is exactly as in the Euclidean case.

7. Invariant factors

We have now proved that any $s \times t$ matrix over a PID R is equivalent to a matrix $\text{diag}(d_1, \ldots, d_u)$ where $d_1 | \cdots | d_u$. It turns out that the diagonal elements $d_1, \ldots, d_u$ are very nearly uniquely determined by A; in fact, they are determined up to associates. We now wish to prove this, and begin by introducing a definition which may appear rather forbidding at first sight.

7.13. Definition. Let A be any $s \times t$ matrix over R, and let $1 \leqslant i \leqslant \min\{s, t\}$. We define $J_i(A)$ to be the ideal of R generated by all the i-minors of A.

An *i-minor* of a matrix A is the determinant of an $i \times i$ submatrix obtained from A by deleting a suitable number of rows and columns (leaving the order of the remaining rows and columns unchanged). Thus an i-minor is an element of the ring from which the entries of the matrix are drawn. We should warn readers that many authors use the word 'minor' to describe the submatrix itself rather than its determinant.

The uniqueness statement which we wish to prove is a consequence of the following:

7.14. Lemma. *Let A, B be $s \times t$ matrices over R, and suppose that A and B are equivalent over R. Then $J_i(A) = J_i(B)$ for $1 \leq i \leq \min\{s,t\}$.*

Before proving this lemma, we justify our interest in it by deducing from it the uniqueness statement we want.

7.15. Theorem. *Let A be an $s \times t$ matrix over R, and let $D = \mathrm{diag}(d_1,\ldots,d_u)$ and $D' = \mathrm{diag}(d'_1,\ldots,d'_u)$ be matrices equivalent to A over R, where $u = \min\{s,t\}$ and $d_1|\cdots|d_u$, $d'_1|\cdots|d'_u$. Then $d_i \sim d'_i$ for $1 \leq i \leq u$.*

Proof. Any non-zero i-minor of D has the form $d_{j_1} d_{j_2} \ldots d_{j_i}$ with $j_1 < j_2 < \cdots < j_i$. These elements are all divisible by the i-minor $d_1 d_2 \ldots d_i$. Hence $J_i(D) = R(d_1 \ldots d_i)$; similarly $J_i(D') = R(d'_1 \ldots d'_i)$. Now D and D' are equivalent to A and so to each other; therefore $J_i(D) = J_i(D')$ by 7.14. Hence by 4.4 we have

$$d_1 \ldots d_i \sim d'_1 \ldots d'_i \quad \text{for } i = 1, 2, \ldots, u. \tag{9}$$

Set $e_0 = 1$, and $e_i = d_1 \ldots d_i$ for $1 \leq i \leq u$. Define e'_i similarly. Then by (9) we have $e_i = v_i e'_i$ ($0 \leq i \leq u$) for some unit $v_i \in R$. Therefore, if $0 \leq i < u$, we have

$$e_{i+1} = d_{i+1} e_i = d_{i+1} v_i e'_i$$

and also

$$e_{i+1} = v_{i+1} e'_{i+1} = v_{i+1} d'_{i+1} e'_i,$$

whence $d_{i+1} v_i = v_{i+1} d'_{i+1}$ and $d_{i+1} = v_i^{-1} v_{i+1} d'_{i+1}$. Hence $d_{i+1} \sim d'_{i+1}$.

To prove the lemma, we begin by making the following observation. Let D be any $m \times m$ matrix over R, and write $D = (\mathbf{d}_1,\ldots,\mathbf{d}_m)$, where $\mathbf{d}_1,\ldots,\mathbf{d}_m$ denote the columns of D. Let $\mathbf{d}'_1$, $\mathbf{d}''_1$ be two other

SUBMODULES OF FREE MODULES

column vectors of length m with entries in R, that is, $m \times 1$ matrices over R. Then

$$\det(\mathbf{d}'_1 + \mathbf{d}''_1, \mathbf{d}_2, \ldots, \mathbf{d}_m) = \det(\mathbf{d}'_1, \mathbf{d}_2, \ldots, \mathbf{d}_m) + \det(\mathbf{d}''_1, \mathbf{d}_2, \ldots, \mathbf{d}_m)$$

and

$$\det(r\mathbf{d}_1, \mathbf{d}_2, \ldots, \mathbf{d}_m) = r\det(\mathbf{d}_1, \mathbf{d}_2, \ldots, \mathbf{d}_m) \quad \text{if } r \in R.$$

These facts are probably familiar to the reader for the case of matrices over a field, and, as in that case, they can be proved either by expanding the determinants by the first column or directly from the definition. Similar remarks apply to other columns. Thus, if we have some $m \times m$ matrix D and replace its i-th column by some R-linear combination of columns $\mathbf{e}_1, \ldots, \mathbf{e}_k$, then the determinant of the resulting matrix is a certain R-linear combination of the determinants of the matrices obtained from D by replacing its i-th column by $\mathbf{e}_1, \ldots, \mathbf{e}_k$ in turn. By applying this principle to each column in succession, we can reach the following conclusion. Suppose we are given some collection $\mathbf{c}_1, \ldots, \mathbf{c}_l$ of column vectors of length m with entries in R. Let $\mathscr{S}$ be the set of all $m \times m$ matrices which can be formed from these columns (with repeats allowed). Now let C be any $m \times m$ matrix whose columns are *linear combinations* of $\mathbf{c}_1, \ldots, \mathbf{c}_l$. Then $\det C$ is a linear combination of elements $\det W$ for $W \in \mathscr{S}$. Therefore $\det C$ belongs to the ideal of R generated by the elements $\det W$ as W runs over $\mathscr{S}$.

Now let $A = (\mathbf{a}_1, \ldots, \mathbf{a}_t)$ be any $s \times t$ matrix over R, and let X be any $t \times t$ matrix over R. Consider AX. The i-th column of this matrix is occupied by the column vector $\sum_{j=1}^{t} x_{ji} \mathbf{a}_j$. We wish to examine a typical $i \times i$ submatrix E of AX. Let $J = \{j_1, \ldots, j_i\}$ be the collection of rows involved in E, written down in natural order. Then the columns of E are linear combinations of 'partial columns' of A, that is, of columns $\mathbf{a}_k^J$ got by selecting the $j_1, \ldots, j_i$-th entries of $\mathbf{a}_k$. Therefore the corresponding i-minor $\det E$ is an R-linear combination of elements

$$\det(\mathbf{a}_{k_1}^J, \ldots, \mathbf{a}_{k_i}^J), \tag{10}$$

determinants of $i \times i$ submatrices formed from selections of the columns $\mathbf{a}_{k_l}^J$. Now the determinant (10) is zero unless all of $k_1, \ldots, k_i$ are distinct, and in the latter case, since we can bring its columns into the same order as that in which they occur in A by a change of sign, it is plus or minus an i-minor of A. Therefore, if E is any

$i \times i$ submatrix of AX, then $\det E$ is a linear combination of i-minors of A and so belongs to the ideal $J_i(A)$ generated by these minors. Therefore
$$J_i(AX) \subseteq J_i(A).$$
By similar arguments with rows we find
$$J_i(YA) \subseteq J_i(A)$$
for any $s \times s$ matrix Y. Therefore
$$J_i(YAX) \subseteq J_i(A).$$
If Y and X are actually invertible and $B = YAX$, then $A = Y^{-1}BX^{-1}$, and so we also have
$$J_i(A) \subseteq J_i(B).$$
Hence these two ideals are equal and this proves Lemma 7.14.

Remark. According to our usual practice we have only stated Lemma 7.14 for PID's. However, the argument shows that it is valid for any commutative ring with 1.

7.16. Definition. Let A be an $s \times t$ matrix over R, and let $D = \text{diag}(d_1, \ldots, d_u)$ be a diagonal matrix equivalent to A over R with $d_1 | d_2 \ldots | d_u$. Then the sequence $d_1, \ldots, d_u$ is called a *sequence of invariant factors* of A over R. $\text{diag}(d_1, \ldots, d_u)$ is called an *invariant factor matrix* for A.

Theorems 7.10 and 7.15 may now be summarized in the following way:

7.17. Theorem. *Two $s \times t$ matrices over a PID R are equivalent over R if and only if they have (to within associates) the same sequence of invariant factors over R.*

Remark. The concept of equivalence, like many others that we have met, is defined relative to a particular ring R. It can happen that two matrices over a ring R, while not equivalent over R itself, become equivalent over some larger ring S (cf. Exercise 3).

8. Summary and a worked example

At this stage the reader might appreciate a short survey of what we have achieved in this rather long chapter. Our stated aim was to study the relationship between a free module F of finite rank and a

SUBMODULES OF FREE MODULES

submodule N, and to prove that a basis $\{f_1, \ldots, f_s\}$ could be chosen for F so that $\{d_1 f_1, \ldots, d_s f_s\}$ was a basis of N for a suitable set of elements $\{d_1, \ldots, d_s\}$ each dividing the next. We showed *en route* that the rank of a free module was a well-defined invariant (7.2), and that the submodule N was itself free (7.8), so that it made sense to talk of a basis for N.

With a given pair of bases **n** and **f** of N and F respectively we associated a matrix A. By studying the relationship between module endomorphisms and matrices we were able to describe the matrices corresponding to a change of basis. We then knew that the matrix associated with two new bases **n*** and **f*** of N and F had the form XAY with X and Y invertible matrices, i.e. the form of a matrix equivalent to A. The original problem could now be translated into a problem about matrices; we simply had to show A was equivalent to $\mathrm{diag}(d_1, \ldots, d_u)$.

Armed with a list of elementary operations, each corresponding to pre- or post-multiplication by an invertible matrix and therefore, when applied to a matrix, producing an equivalent one, we set about reducing A to the desired diagonal form. In the case of an ED it was easy to show, with the help of the Euclidean function ϕ, that this reduction could be carried out in a finite number of steps. To achieve this for an arbitrary PID we had to find a substitute for ϕ; here our work in Chapter 4 on uniqueness of factorization in a PID came to the rescue, enabling us to define a length function on the ring. With the help of an additional operation we then carried through a slightly modified argument to complete the proof in the general case. Finally, we showed the elements $d_1 | d_2 | \cdots | d_u$ were uniquely determined up to associates.

We conclude the chapter by giving some numerical examples to illustrate how the reduction procedure given in §5 works in practice.

Worked Example. Let
$$T = \begin{bmatrix} 7 & 8 & 9 \\ 4 & 5 & 6 \\ 1 & 2 & 3 \end{bmatrix}.$$

Find invertible 3×3 matrices X and Y over **Z** such that XTY is an invariant-factor matrix for T.

The first task is to reduce T to an invariant-factor matrix by elementary row and column operations. Since we have to find the matrices X and Y, we have to keep track of the operations used. We give some shorthand notation to describe these operations.

$R_i \leftrightarrow R_j$ means interchange the i-th and j-th rows,
$R_i + cR_j$ means add c times row j to row i,
uR_i means multiply row i by the unit u (in the present case u will be ± 1).

These all denote elementary row operations, and we use the corresponding notation for elementary column operations with R replaced by C.

It is always quicker in practice (though not strictly necessary) to begin the reduction procedure by bringing a non-zero element of smallest ϕ-value to the leading position. So we have

$$\begin{bmatrix} 7 & 8 & 9 \\ 4 & 5 & 6 \\ 1 & 2 & 3 \end{bmatrix} \to R_3 \leftrightarrow R_1 \begin{bmatrix} 1 & 2 & 3 \\ 4 & 5 & 6 \\ 7 & 8 & 9 \end{bmatrix}$$

$$\to \begin{matrix} R_2 - 4R_1 \\ R_3 - 7R_1 \end{matrix} \begin{bmatrix} 1 & 2 & 3 \\ 0 & -3 & -6 \\ 0 & -6 & -12 \end{bmatrix}$$

$$\to \begin{matrix} C_2 - 2C_1 \\ C_3 - 3C_1 \end{matrix} \begin{bmatrix} 1 & 0 & 0 \\ 0 & -3 & -6 \\ 0 & -6 & -12 \end{bmatrix}$$

which brings us to (£). As the first row and column now remain unchanged, we can suppress them provided we continue to number the rows and columns as rows and columns of the original matrix. So we continue as follows:

$$\begin{bmatrix} -3 & -6 \\ -6 & -12 \end{bmatrix} \to R_3 - 2R_2 \begin{bmatrix} -3 & -6 \\ 0 & 0 \end{bmatrix} \to \begin{matrix} C_3 - 2C_2 \\ -1 \times C_2 \end{matrix} \begin{bmatrix} 3 & 0 \\ 0 & 0 \end{bmatrix}.$$

Thus we have reduced our matrix T to

$$\begin{bmatrix} 1 & 0 & 0 \\ 0 & 3 & 0 \\ 0 & 0 & 0 \end{bmatrix}$$

and 1, 3, 0 is a sequence of invariant factors for T over **Z**.

SUBMODULES OF FREE MODULES

To find the matrix X we apply the elementary row operations used above to the 3×3 identity matrix 1_3, and to find Y we apply the column operations to 1_3. It is easy to confirm that this gives

$$X = \begin{bmatrix} 0 & 0 & 1 \\ 0 & 1 & -4 \\ 1 & -2 & 1 \end{bmatrix} \qquad Y = \begin{bmatrix} 1 & 2 & 1 \\ 0 & -1 & -2 \\ 0 & 0 & 1 \end{bmatrix}$$

and it is a useful check on the calculations to work out the product XTY directly.

An easy situation in which case 3 arises is obtained by beginning with the matrix

$$\begin{bmatrix} 2 & 0 \\ 0 & 3 \end{bmatrix}.$$

In that case a reduction procedure is

$$\begin{bmatrix} 2 & 0 \\ 0 & 3 \end{bmatrix} \xrightarrow{R_1 + R_2} \begin{bmatrix} 2 & 3 \\ 0 & 3 \end{bmatrix} \xrightarrow{\substack{C_2 - C_1 \\ C_1 \leftrightarrow C_2}} \begin{bmatrix} 1 & 2 \\ 3 & 0 \end{bmatrix}$$

$$\xrightarrow{\substack{C_2 - 2C_1 \\ R_2 - 3R_1 \\ -1 \times R_2}} \begin{bmatrix} 1 & 0 \\ 0 & 6 \end{bmatrix}.$$

Therefore 1, 6 is the sequence of invariant factors in this case.

Exercises for Chapter 7

1. Let $A = \begin{bmatrix} -4 & -6 & 7 \\ 2 & 2 & 4 \\ 6 & 6 & 15 \end{bmatrix}$.

   Find matrices X and Y over $\mathbf{Z}$ such that XAY is an invariant factor matrix for A over $\mathbf{Z}$.

2. Calculate invariant factor matrices over $\mathbf{Q}[x]$ for

   (i) $\begin{bmatrix} x & 0 & 0 \\ 0 & 1-x & 0 \\ 0 & 0 & 1-x^2 \end{bmatrix}$, (ii) $\begin{bmatrix} 1-x & 1+x & x \\ x & 1-x & 1 \\ 1+x & 2x & 1 \end{bmatrix}$.

3. What does the relation of equivalence mean for 1×1 matrices? Give an example of two matrices over $\mathbf{Z}$ which, though not equivalent over $\mathbf{Z}$, are equivalent over $\mathbf{Q}$.

4. Show that every $s \times t$ matrix over a field **k** is equivalent over **k** to a matrix $\operatorname{diag}(1,1,\ldots,1,0,\ldots,0)$. Into how many equivalence classes under the relation of equivalence over **k** does the set of all $s \times t$ matrices over **k** fall?

5. Let R be a PID and let A be a $n \times n$ matrix over R. Show that A is invertible if and only if A is equivalent to the $n \times n$ identity matrix over R. Show that, if R is a Euclidean domain, the $n \times n$ elementary matrices generate the group of all $n \times n$ invertible matrices over R. What is the corresponding result when R is an arbitrary PID?

6. Let A be the matrix
$$\begin{bmatrix} 2 & 1+i & 1-i \\ 8+6i & -4 & 0 \end{bmatrix}$$
with entries in the ring R of Gaussian integers. Find square matrices X and Y over R such that XAY is an invariant factor matrix for A over R. What is the answer when R is replaced by **C**?

7. Let R be an integral domain. Prove that no subset of $(_R R)^n$ which is linearly independent over R can contain more than n elements. (*Hint*: embed R in its field of fractions **k** constructed as in Chapter 4, §1, and regard $(_R R)^n$ as embedded in $\mathbf{k}^n$.)

8*. Let F be a free module over a PID R with a basis $\{f_1,\ldots,f_n\}$. Let $r_1,\ldots,r_n$ be n elements of R, and suppose that their hcf is [1]. Let $f = \sum_{i=1}^n r_i f_i$. Prove (using 7.1) that $F = Rf \oplus F^*$ for some submodule F^* of F. Deduce that there is an invertible $n \times n$ matrix over R with first column $r_1 \ldots, r_n$.

9*. Show that
$$A = \begin{bmatrix} 2x & 0 \\ x & 2 \end{bmatrix}$$
is not equivalent over $\mathbf{Z}[x]$ to a diagonal matrix. (*Hint*: consider the ideals $J_i(A)$.)

CHAPTER EIGHT

Decomposition theorems

We are now in a position to state and prove the main theorem of this book, which is a theorem giving detailed information about the structure of finitely-generated modules over a principal ideal domain R. It leads, in fact, to a classification of such modules (in terms of certain sequences of elements of R), achieved by expressing them as direct sums of certain cyclic submodules. Later in the chapter we show how to refine such a direct sum to its most basic form where the components can be decomposed no further. At each stage we take a careful look at the uniqueness of the various decompositions derived.

1. The main theorem

First we need an elementary lemma about direct sums of modules over any ring with 1.

8.1. Lemma. *Let L be a module over a ring R, and suppose L is an internal direct sum $L = L_1 \oplus \cdots \oplus L_t$ of submodules L_i. For each i let N_i be a submodule of L_i, and let $N = \sum_{i=1}^{t} N_i$. Then, if v is the natural homomorphism $L \to L/N$, we have $L/N = v(L) = v(L_1) \oplus \cdots \oplus v(L_t)$ and $v(L_i) \cong L_i/N_i$.*

Proof. If $l \in L$, then $l = \sum_{i=1}^{t} l_i$ with $l_i \in L_i$, and so $v(l) = \sum_{i=1}^{t} v(l_i) \in \sum_{i=1}^{t} v(L_i)$. Therefore $v(L) = \sum_{i=1}^{t} v(L_i)$. To see that the sum is direct let $x \in v(L_i) \cap \sum_{j \neq i} v(L_j)$. Then $x = v(l'_i) = \sum_{j \neq i} v(l'_j)$ with $l'_k \in L_k$. Hence $0 = v(l'_i - \sum_{j \neq i} l'_j)$ and so $l'_i - \sum_{j \neq i} l'_j \in \ker v = N = \sum N_i$. Therefore $l'_i - \sum_{j \neq i} l'_j = \sum_{k=1}^{t} n_k$ with $n_k \in N_k$. Since the sum

$\sum L_i$ is direct, we find $l'_i = n_i \in N_i \subseteq \ker \nu$, and so $x = \nu(l'_i) = 0$. Hence $\nu(L) = \nu(L_1) \oplus \cdots \oplus \nu(L_t)$. Now the kernel of ν restricted to L_i is $N \cap L_i = N_i$, and so by 5.10 we have $\nu(L_i) \cong L_i/N_i$.

We now go on to our main result.

8.2. Theorem. *Let R be a PID and M be an FG R-module. Then M can be expressed as an (internal) direct sum*

$$M = M_1 \oplus \cdots \oplus M_s \quad (s \geqslant 0)$$

where

(a) *M_i is a non-trivial cyclic submodule of M of order d_i, and*
(b) *$d_1 | d_2 | \cdots | d_s$.*

Remarks. 1. We recall that according to our conventions the zero module is a direct sum of the empty set of submodules.
2. We have so far only defined the order *ideal* $\mathbf{o}(N)$ of a cyclic R-module N. However, if R is a PID, then $\mathbf{o}(N)$ is a principal ideal and so has the form dR for some $d \in R$. By 4.4 d is determined up to unit factor and is called an *order* of N. We say that N is *of order d*. If $N = Rn$ and $r \in R$, then $rn = 0 \Leftrightarrow r \in \mathbf{o}(N) \Leftrightarrow d | r$. As we have pointed out, the order of a finite cyclic group in the usual sense of group theory is the positive generator of its order ideal and so is an order in the sense we have just described.
3. Let $Z = Rz$ be a cyclic R-module with order ideal $J = dR$. Then $Z = \{0\} \Leftrightarrow J = R \Leftrightarrow d$ is a unit. Thus the statement that the M_i are all non-trivial is equivalent to the statement that no d_i is a unit.
4. The condition $d_1 | \cdots | d_s$ is equivalent to $\mathbf{o}(M_1) \supseteq \cdots \supseteq \mathbf{o}(M_s)$.

Proof of the theorem. Let M be an FG R-module. Then by 6.10 there exists an epimorphism $\phi: F \to M$ where F is a free R-module of finite rank t say. Let $N = \ker \phi$. Then by 5.9 there exists an isomorphism $\psi: F/N \to M$ making the diagram

DECOMPOSITION THEOREMS

commute, where ν is the natural homomorphism. Now by 7.1 there exists a basis $\{f_1,\ldots, f_t\}$ for F and elements $c_1|c_2|\cdots|c_t$ in R such that the elements $c_1 f_1,\ldots, c_t f_t$ generate N. Therefore $F = Rf_1 \oplus \cdots \oplus Rf_t$ and $N = R(c_1 f_1) \oplus \cdots \oplus R(c_t f_t)$ – here some of the elements $c_i f_i$ may be zero. By 8.1 F/N is the direct sum of its cyclic submodules $\nu(Rf_i) = R\nu(f_i)$. Now $\nu(f_i)$ has order c_i, since if $r \in R$ then $r\nu(f_i) = 0 \Leftrightarrow \nu(rf_i) = 0 \Leftrightarrow rf_i \in N \Leftrightarrow c_i|r$. Hence we have

$$F/N = R\nu(f_1) \oplus \cdots \oplus R\nu(f_t), \tag{1}$$

where $\nu(f_i)$ has order c_i and $c_1|c_2|\cdots|c_t$. Since ψ is an isomorphism, it maps the direct decomposition (1) of F/N into a direct decomposition of M. We now want to omit the trivial summands. Let u be the last integer i such that c_i is a unit. Then $c_1,\ldots, c_u$ are all units by the divisibility condition, and the corresponding modules in (1) are exactly the zero modules and can be omitted. Therefore, if $s = t - u$, then

$$M = M_1 \oplus \cdots \oplus M_s,$$

where $M_i = R\psi\nu(f_{u+i}) = R\phi(f_{u+i})$ is a non-trivial cyclic module of order $d_i = c_{u+i}$ and $d_1|d_2|\cdots|d_s$. This concludes the proof.

8.3. Corollary. *With the hypotheses of Theorem 8.2, $M = T \oplus F$, where T is the torsion submodule of M and F is a free submodule of finite rank.*

Proof. We already know from 6.5 that the set T of torsion elements of M is a submodule of M – this is what we mean by the torsion submodule of M. In the decomposition of 8.2 for M, let $l + 1$ be the first integer j such that $d_j = 0$. Then $d_{l+1} = \cdots = d_s = 0$ by 8.2(b), and so each of $M_{l+1},\ldots, M_s$ is a torsion-free cyclic module. Hence by 6.8 $F = M_{l+1} \oplus \cdots \oplus M_s$ is free of rank $s - l$. Let $T^* = M_1 \oplus \cdots \oplus M_l$. Then clearly $M = T^* \oplus F$, and we claim that in fact $T^* = T$. For let $m \in T^*$. Then $m = m_1 + \cdots + m_l$ with $m_i \in M_i$, and since d_i divides d_l for $i \leqslant l$, we have $d_l m = d_l m_1 + \cdots + d_l m_l = 0$. Since $d_l \neq 0$, this shows that every element of T^* is a torsion element, that is, $T^* \subseteq T$. On the other hand, let n be any torsion element of M. Then, since $M = T^* \oplus F$, we have $n = t + f$ with

$t \in T^*, f \in F$. Since n is a torsion element, $rn = 0$ for some $0 \neq r \in R$. Hence $rt + rf = 0$. Since the sum $T^* \oplus F$ is direct, this gives $rf = 0$. But F is torsion-free and so $f = 0$. Hence $n \in T^*$, and therefore $T^* = T$, as claimed.

Remark. The submodule F in 8.3 is not in general uniquely determined. In the **Z**-module $M = \mathbf{Z}_2 \oplus \mathbf{Z}$, the torsion submodule T is the cyclic submodule generated by $(1,0)$. The reader can easily verify that the two elements $(0,1)$ and $(1,1)$ generate cyclic submodules F and F^* such that $M = T \oplus F = T \oplus F^*$, and that $F \neq F^*$.

8.4. Corollary. *An FG torsion-free module over a PID R is free.*

Proof. In the notation of 8.3, if M is torsion-free, then we must have $T = \{0\}$, and so $M = F$ which is free.

Examples. It is obviously of interest to know whether a theorem remains true under weaker hypotheses than those under which it is stated. We shall now show that the two main hypotheses of 8.2 cannot be omitted.
1. The hypothesis that R is a PID is not redundant. For let $R = \mathbf{Z}[x]$. Then no linearly independent set in $_RR$ contains more than one element. For, if a and b are two non-zero elements of R, then we have $0 = ba - ab$, which is a non-trivial linear relation between a and b. Let $J = 2R + xR$, the ideal of R generated by 2 and x. Then J, as a submodule of $_RR$, can be generated by two elements. Now by Chapter 4, Exercise 8, J is not a principal ideal, that is, J is not a cyclic submodule of $_RR$. Therefore, if J were a direct sum of cyclic modules, these modules would all be torsion-free (since $_RR$ is) and there would be at least two of them. The generators of two of these modules would then be linearly independent, which we have just seen is impossible. (In this argument $\mathbf{Z}[x]$ can be replaced by any integral domain which is not a PID.)
2. The hypothesis that M is FG obviously cannot be removed, since any module which is a direct sum of finitely many cyclic submodules must be FG. Since we have not discussed infinite direct sums, we cannot pursue this question any further here.

DECOMPOSITION THEOREMS

2. Uniqueness of the decomposition

What does it mean to ask whether a direct decomposition of a module is unique? For the type of decomposition described in 8.2 this question might be expressed in its most uncompromising form as follows:

'Does the existence of two decompositions of the form

$$M = M_1 \oplus \cdots \oplus M_r = M_1' \oplus \cdots \oplus M_s'$$

with M_i, M_i' non-zero cyclic submodules of M such that $o(M_1) \supseteq \cdots \supseteq o(M_r)$ and $o(M_1') \supseteq \cdots \supseteq o(M_s')$ always imply that $r = s$ and $M_i = M_i'$ for $i = 1, \ldots, r$?'

The simple answer to this question is no. For given one such decomposition in which several components have the same order ideal we can obviously get another decomposition of the same type simply by permuting the components. For example, if $M = Z_1 \oplus Z_2$ is the internal direct sum of two copies Z_i of the Z-module Z, then $M = Z_2 \oplus Z_1$ is another decomposition; both have the form described in 8.2.

Even if we weaken the uniqueness requirement slightly, and replace the unrealistic demand $M_i = M_i'$ by the more modest proposal that the M_i should be equal to the M_j' *in some order*, we still get a negative answer to the question, and for an elementary reason. This is because one can choose a basis for a free module in a variety of ways. For example, $\mathbf{Z} \oplus \mathbf{Z} = \mathbf{Z}(1,0) \oplus \mathbf{Z}(0,1)$ is a decomposition of the type of 8.2 for the external direct sum of Z with itself. But as we saw in Chapter 7, §2, for any Z-matrix

$$\begin{bmatrix} a & b \\ c & d \end{bmatrix}$$

with determinant ± 1 the direct decomposition $\mathbf{Z}(a,c) \oplus \mathbf{Z}(b,d)$ is another of the required form, the order ideals of both cyclic summands being zero.

Even for torsion modules the answer to our question is still negative. For let $M = A \oplus B$ be the internal direct sum of two cyclic groups $A = \mathbf{Z}a$ and $B = \mathbf{Z}b$ of order two, considered as a Z-module in the usual way. This is a decomposition of type 8.2 because $o(A) = o(B) = 2\mathbf{Z}$. But $M = \mathbf{Z}a \oplus \mathbf{Z}(a+b)$ is another decomposition of this type neither of whose direct summands is equal to B. Thus there is no hope of salvaging this simple-minded concept of uniqueness for the decompositions described in 8.2.

All is not lost, however. The more humble question 'What degree of uniqueness, if any, does our decomposition have?' does have some positive, and indeed very useful, answers. For example, the number of summands in such a decomposition is an invariant of the module; so is the nested sequence $\{o(M_i)\}$ of order ideals which occurs in 8.2. We shall prove

8.5. Theorem. *If R is a PID and $M = M_1 \oplus \cdots \oplus M_s = M'_1 \oplus \cdots \oplus M'_t$ are two direct decompositions of the R-module M, where M_i is a non-trivial cyclic module of order d_i, M'_j is a non-trivial cyclic module of order d'_j and $d_1|d_2|\cdots|d_s$, $d'_1|d'_2|\cdots|d'_t$, then $s = t$ and $Rd_i = Rd'_i$ for $i = 1, 2, \ldots, s$. In particular d_i and d'_i are associates.*

On the basis of this theorem we make some definitions.

8.6. Definitions. Let M be a FG R-module over a PID R. Then by 8.2 we can write $M = M_1 \oplus \cdots \oplus M_s$, where M_i is a non-trivial cyclic module of order d_i, and $d_1|d_2|\cdots|d_s$. By 8.5 the sequence $d_1, d_2, \ldots, d_s$ is determined up to unit factors. It is called a *sequence of invariant factors* of M. Notice that, while the invariant factors of a matrix may be units, by this definition the invariant factors of a module are always non-units.

Let $l+1$ be the first integer i such that $d_i = 0$. Then $d_{l+1} = \cdots = d_s = 0$, and 8.5 tells us that the integers l and $s-l$ are uniquely determined by the module M. We call $s-l$ the *torsion-free rank* of M. The ordered set $d_1, \ldots, d_l$ is called a *sequence of torsion invariants* of M. Clearly one can derive a sequence of invariant factors of a module from knowledge of its torsion invariants and its torsion-free rank.

Remarks. 1. With the above notation we have $M = T \oplus F$, where F is free of rank $s-l$, and $T = M_1 \oplus \cdots \oplus M_l$ is the torsion submodule of M; this follows by the argument of 8.3. Hence $M/T = F \oplus T/T \cong F/F \cap T = F/\{0\} \cong F$. Therefore in any decomposition of M as in 8.5 the number of torsion-free cyclic summands is the rank of the free module M/T and so is uniquely determined by M. The torsion-free rank of M is simply the rank of the free module M/T.

DECOMPOSITION THEOREMS

2. Theorems 8.1 and 8.5 give us a *classification* of FG modules over a PID R. By these theorems, each such module M determines an integer $s \geqslant 0$ and a sequence

$$[d_1]|[d_2]|\cdots|[d_s] \qquad (2)$$

of classes of associates in R, where each d_i is a non-unit of R. If two modules M and M' are isomorphic, then by 8.5 they determine the same sequence. Conversely, if M and M' correspond to the same sequence (2), then $M = M_1 \oplus \cdots \oplus M_s$, $M' = M'_1 \oplus \cdots \oplus M'_s$, where M_i and M'_i are cyclic of order d_i. Therefore by 6.11 $M_i \cong M'_i$, and it follows that $M \cong M'$. Finally we observe that every sequence of the form (2) corresponds to a module, namely the external direct sum of cyclic modules of orders $d_1, \ldots, d_s$. Of course $_R(R/Rd_i)$ is an example of a cyclic module of order d_i. Thus isomorphism classes of FG R-modules are in one-one correspondence with sequences of the form (2).

Now we must prove Theorem 8.5; we do so by deducing it from the theorem which gives the uniqueness of the invariant factors of a matrix. We have already seen that if $\epsilon: F \to M$ is an epimorphism with kernel N, where F is free, then certain choices of bases in F and N determine decompositions of M as a direct sum of cyclic submodules. Our proof of 8.5 is carried out by showing that, conversely, certain direct decompositions of M determine bases in F and N of the type discussed in 7.1. The following lemma will be useful.

8.7. Lemma. *Let M be an R-module. Let x and y be elements of M, and suppose that x has order $d \neq 0$. Suppose further that $Ry \supseteq Rx$ and $dy = 0$. Then $Rx = Ry$.*

Proof. By assumption we have $x = ry$ for some $r \in R$. Let h be an hcf of r and d. Then $r = r_1 h$, $d = d_1 h$ for some $r_1, d_1 \in R$. Then $x = ry = r_1 hy$ and so $d_1 x = r_1 d_1 hy = r_1 dy = 0$. Therefore since x has order d, $d|d_1$. Hence $d \sim d_1$ and h is a unit. Therefore $(r, d) = [1]$, and so $\exists u, v \in R$ such that $ur + vd = 1$. Hence $y = (ur + vd)y = ux \in Rx$, as required.

8.8. Lemma. *Let $M = Rx_1 \oplus \cdots \oplus Rx_t$ be a direct sum of non-trivial cyclic torsion submodules Rx_i of order $d_i \neq 0$, where $d_t|\cdots|d_1$. Let $\epsilon: F \to M$ be an epimorphism, where F is free of finite rank s.*

Then there exists a basis $\{f_1, \ldots, f_s\}$ of F and elements $y_i \in Rx_i$ such that

(i) $M = Ry_1 \oplus \cdots \oplus Ry_t$ *and y_i has order d_i, and*
(ii) $\epsilon(f_i) = y_i$ *for $1 \leq i \leq t$, $\epsilon(f_i) = 0$ for $t < i \leq s$.*

Proof. Notice that we have written the cyclic submodules in the opposite order to the usual one; this turns out to simplify the notation. The proof is by induction on t. If $t = 0$, then $M = \{0\}$, ϵ is the zero map, and any basis of F satisfies the conditions.

Now assume that $t > 0$ and that the theorem holds for direct sums with fewer than t summands. Now since ϵ is surjective, we have $\epsilon(g_1) = x_1$ for some $g_1 \in F$; and since $x_1 \neq 0$, we must have $g_1 \neq 0$. Applying 7.1 to the submodule Rg_1 of F, we find that there exists a basis $\{f_1', \ldots, f_s'\}$ of F and a basis element g_1' of Rg_1 such that $g_1' = rf_1'$ for some $r \in R$. Then $g_1 = ug_1'$ for some unit u, and so $g_1 = urf_1'$. Let $y_1 = \epsilon(f_1')$. Then $x_1 = \epsilon(g_1) = \epsilon(urf_1') = ury_1$, and so $x_1 \in Ry_1$. Now $d_1 y_1 = 0$. This is because the order d_i of each x_i divides d_1, so that $d_1 x_i = 0$ for each i and $d_1 M = \{0\}$. It now follows from 8.7 that $Rx_1 = Ry_1$. Since any two generators of a cyclic module have the same order, y_1 must have order d_1. Furthermore

$$M = Ry_1 \oplus M_1 \qquad (3)$$

where $M_1 = Rx_2 \oplus \cdots \oplus Rx_t$.

Let π be the projection of M on M_1 associated with the decomposition (3). Thus, if $m = ry_1 + m_1$ with $m_1 \in M_1$ and $r \in R$, then $\pi(m) = m_1$. Let $F_1 = Rf_2' \oplus \cdots \oplus Rf_s'$. Then $F = Rf_1' \oplus F_1$. Now since $\pi: M \to M_1$ and $\epsilon: F \to M$ are epimorphisms, $\pi\epsilon$ is an epimorphism from F to M_1. If $f = xf_1' + f^* \in F$ with $f^* \in F_1$ and $x \in R$, then as $\pi\epsilon(f_1') = \pi(y_1) = 0$ we have $\pi\epsilon(f) = \pi\epsilon(f^*)$. Therefore the restriction of $\pi\epsilon$ to F_1 is an epimorphism. It now follows from our inductive hypothesis that there exists a basis $\{f_2^*, \ldots, f_s^*\}$ of F_1 and elements $y_i \in Rx_i$ of order d_i for $2 \leq i \leq t$ such that $\pi\epsilon(f_i^*) = y_i$ ($2 \leq i \leq t$), and $\pi\epsilon(f_i^*) = 0$ if $t < i \leq s$. This means that $\epsilon(f_i^*) = y_i + r_i y_1$ ($2 \leq i \leq t$) and $\epsilon(f_i^*) = r_i y_1$ ($t < i \leq s$) for suitable $r_i \in R$. Let

$$f_1 = f_1'$$

and

$$f_i = f_i^* - r_i f_1' \quad (i \neq 1).$$

DECOMPOSITION THEOREMS

Now $\{f'_1, f^*_2, \ldots, f^*_s\}$ is certainly a basis of F and hence so is $\{f_1, \ldots, f_s\}$, since the matrix relating these two sets of elements has determinant 1. Further,

$$\epsilon(f_1) = \epsilon(f'_1) = y_1, \; \epsilon(f_i) = \epsilon(f^*_i - r_i f'_i) = y_i + r_i y_1 - r_i y_1 = y_i$$

if $2 \leq i \leq t$, and $\epsilon(f_i) = 0$ if $t < i \leq s$. This concludes the proof.

8.9. Corollary. *With the notation of 8.8, assume further that M is a torsion module and $s > 0$. Let $N = \ker \epsilon$. Then there is a basis of N which has matrix* $\mathrm{diag}(1, \ldots, 1, d_t, \ldots, d_1)$, *with $s - t$ 1's, with respect to a basis of F.*

Proof. We show that

$$N = R(d_1 f_1) \oplus \cdots \oplus R(d_t f_t) \oplus Rf_{t+1} \oplus \cdots \oplus Rf_s. \quad (4)$$

Indeed, if $f \in N$, since $\{f_1, \ldots, f_s\}$ is a basis of F we have $f = \sum_{i=1}^{s} r_i f_i$ for suitable $r_i \in R$. Then $0 = \epsilon(f) = \sum_{i=1}^{s} r_i \epsilon(f_i) = \sum_{i=1}^{t} r_i y_i$. Since the sum of the submodules Ry_i is direct, we have $r_i y_i = 0$ for each i, and so the order d_i of y_i divides r_i if $1 \leq i \leq t$. Conversely, this condition clearly ensures $f \in N$, and so (4) is established. Since M is a torsion module, no d_i is zero, and so $\{d_1 f_1, \ldots, d_t f_t, f_{t+1}, \ldots, f_s\}$ is a basis of N. If we write this basis in the opposite order, its matrix with respect to $\{f_s, \ldots, f_1\}$ is the required diagonal matrix.

Proof of Theorem 8.5. This is now fairly straightforward, most of the work having already been done. We have

$$M = M_1 \oplus \cdots \oplus M_s = M'_1 \oplus \cdots \oplus M'_t \quad (5)$$

where M_i, M'_j are non-trivial cyclic modules of orders d_i, d'_j respectively, and $d_1 | d_2 | \cdots | d_s$, $d'_1 | d'_2 | \cdots | d'_t$. Notice that the modules are now numbered in the more usual order. Let $u + 1$ be the first integer i such that $d_i = 0$, and let v be defined similarly with respect to the second decomposition. Then by the argument of 8.3

$$T = M_1 \oplus \cdots \oplus M_u = M'_1 \oplus \cdots \oplus M'_v \quad (6)$$

is the torsion submodule of M, and $M/T \cong M_{u+1} \oplus \cdots \oplus M_s$ is free of rank $s - u$; similarly M/T is free of rank $t - v$. Therefore by 7.2, which gives the invariance of the rank of free modules, we get

$$s - u = t - v. \tag{7}$$

We may suppose without loss of generality that $u \geqslant v$. Now if $u = 0$, then (6) gives $v = 0$, (7) gives $s = t$, and since all the d_i and d'_j are then zero, the result is proved. Therefore we may suppose $u \neq 0$. By 6.10 there exists an epimorphism ϵ of a free module F of rank u onto T, and by 8.9 applied successively to the first and second decompositions of T, each of the two matrices $\mathrm{diag}(d_1,\ldots,d_u)$ and $\mathrm{diag}(1,\ldots,1,d'_1,\ldots,d'_v)$ (with $u - v$ 1's) is the matrix of a basis of $N = \ker \epsilon$ with respect to a basis of F. By the arguments of Chapter 7, §2, these two matrices are equivalent. Therefore by 7.17 $d_i \sim 1$ for $i = 1,\ldots, u - v$. Since by assumption no d_i is a unit, we must have $u = v$; furthermore, 7.17 then gives $d_i \sim d'_i$ for $i = 1,\ldots, u$. Equation (7) now gives $s = t$, and since the remaining d_i and d'_j are all zero, the proof is complete.

We shall give an independent and perhaps rather more elegant approach to Theorems 8.2 and 8.5 in the next chapter.

3. The primary decomposition of a module

It is natural to ask in the light of 8.2 whether the summands M_i there obtained can be decomposed into yet 'smaller' modules. We shall now show that this is sometimes possible. We have already seen that $\mathbf{Z}_6 \cong \mathbf{Z}_3 \oplus \mathbf{Z}_2$ as rings (therefore also as Abelian groups and as $\mathbf{Z}$-modules), and the decomposition we now give is along the lines suggested by this example.

8.10. Lemma. *Let M be a non-trivial R-module, and suppose $dM = 0$, where $d \in R$ and is neither zero nor a unit. Let $d = up_1^{\alpha_1} \ldots p_k^{\alpha_k}$, where u is a unit and the p_i are pairwise non-associated primes in R. Then M can be expressed as a direct sum $M = M_1 \oplus \cdots \oplus M_k$, where $p_i^{\alpha_i} M_i = \{0\}$. The submodules M_i are uniquely determined by these conditions.*

Proof. Notice that since R is a UFD d can certainly be expressed in the stated form; we take some expression $d = \text{unit} \times \text{product of}$

DECOMPOSITION THEOREMS

primes, and at the expense of bringing various units out to the front, we collect together all the primes associated to a given one. Set

$$d_i = d/p_i^{\alpha_i} = u \prod_{j \neq i} p_j^{\alpha_j}$$

We first prove that, if there exists a decomposition

$$M = M_1 \oplus \cdots \oplus M_k \quad \text{with} \quad p_i^{\alpha_i} M_i = \{0\}, \tag{8}$$

then $M_i = d_i M$, thereby showing that the components M_i are uniquely determined in the strongest possible sense. Now clearly we have $d_i M \subseteq d_i M_1 + \cdots + d_i M_k = d_i M_i \subseteq M_i$, because $d_i M_j = 0$ for $j \neq i$ (since $p_j^{\alpha_j} | d_i$). On the other hand, since $(d_i, p_i^{\alpha_i}) = [1]$, there exist $r, s \in R$ such that $rd_i + sp_i^{\alpha_i} = 1$. Therefore, if $m \in M_i$, then $m = (rd_i + sp_i^{\alpha_i})m = d_i(rm) \in d_i M$. Hence $M_i \subseteq d_i M \subseteq d_i M_i \subseteq M_i$, and we have equality throughout.

To show the existence of a decomposition we must therefore take $M_i = d_i M$ and show that (8) holds. Clearly $p_i^{\alpha_i} M_i = \{0\}$, since $d_i p_i^{\alpha_i} = d$. Now since the hcf of $\{d_1, \ldots, d_k\}$ is [1], there exist $r_1, \ldots, r_k \in R$ such that $\sum r_i d_i = 1$. Therefore, if $x \in M$, then $x = 1x = \sum d_i(r_i x) \in \sum d_i M = \sum M_i$. To see that the sum is direct, we notice that any element $y \in \sum_{j \neq i} M_j$ satisfies $d_i y = 0$; for, as pointed out above, if $j \neq i$, we have $d_i M_j = 0$. If y also belongs to M_i, then $p_i^{\alpha_i} y = 0$. Choosing r and s as in the last paragraph we then have $y = (rd_i + sp_i^{\alpha_i})y = 0$. This concludes the proof.

8.11. Corollary. *If $M = Rm$ is cyclic, then $M_i = R(d_i m)$ is cyclic. In this case, if M has order precisely d, then M_i has order $p_i^{\alpha_i}$.*

8.12. Definitions. The modules M_i of 8.10 are called the *primary components* of M. A module N such that $p^\alpha N = \{0\}$ (where p is a prime) is called a *primary module* or a *p-torsion module*. A cyclic *p*-torsion module is called a *primary cyclic module*, a *cyclic module of prime-power order*, or a *cyclic module of p-power order*.

We can use Lemma 8.10 to obtain a refinement of our decomposition theorem 8.2, but before doing so we will state and prove the following 'collecting-up lemma' for direct sums of cyclic modules of relatively prime orders.

8.13. Lemma. *Let $M = M_1 \oplus \cdots \oplus M_k$, where M_i is a cyclic torsion module of order r_i and $(r_i, r_j) = [1]$ for $i \neq j$. Then M is cyclic of order $r_1 \ldots r_k$.*

Proof. Let $M_i = Rm_i$ and $m = m_1 + \cdots + m_k$. Let $d = r_1 \ldots r_k$. Now for $s \in R$ we have $sm = 0 \Leftrightarrow sm_i = 0$ for each $i \Leftrightarrow r_i | s$ for each i. Since the r_i are coprime in pairs this means that d must divide s, and so we find that m has order d. Writing $d_i = d/r_i$, we clearly have $d_i m = d_i m_i$. Since $(r_i, d_i) = [1]$, there exist $t, u \in R$ such that $tr_i + ud_i = 1$, and so $m_i = (tr_i + ud_i)m_i = u(d_i m_i) \in R(d_i m_i) = R(d_i m) \subseteq Rm$. Therefore Rm contains each m_i and must be the whole of M.

Example. Let M be the **Z**-module $\mathbf{Z}_6 \oplus \mathbf{Z}_{20} \oplus \mathbf{Z}_{36}$. Then by 8.10 and 8.11 and with a certain amount of abuse of notation we have $\mathbf{Z}_6 = \mathbf{Z}_3 \oplus \mathbf{Z}_2$, $\mathbf{Z}_{20} = \mathbf{Z}_5 \oplus \mathbf{Z}_4$, $\mathbf{Z}_{36} = \mathbf{Z}_9 \oplus \mathbf{Z}_4$ and so

$$M = (\mathbf{Z}_2 \oplus \mathbf{Z}_4 \oplus \mathbf{Z}_4) \oplus (\mathbf{Z}_3 \oplus \mathbf{Z}_9) \oplus \mathbf{Z}_5.$$

We have bracketed the 2-torsion, 3-torsion and 5-torsion components of M and thus expressed M as a direct sum of its primary components. To obtain a decomposition of M as described in 8.2 we select one cyclic submodule of largest possible order from each primary component and then bracket them together to give the cyclic submodule M_s. We then repeat the process with the submodules remaining to obtain M_{s-1}, and so on, at any given stage ignoring those primary components which have already been exhausted. Thus

$$M = \mathbf{Z}_2 \oplus (\mathbf{Z}_4 \oplus \mathbf{Z}_3) \oplus (\mathbf{Z}_4 \oplus \mathbf{Z}_9 \oplus \mathbf{Z}_5)$$
$$= \mathbf{Z}_2 \oplus \mathbf{Z}_{12} \oplus \mathbf{Z}_{180} \quad \text{by 8.12.}$$

It follows that 2, 12, 180 is a sequence of torsion invariants for M.

8.14. Theorem. *Let M be an FG module over a PID R. Then M can be expressed as a direct sum*

$$M = Z_1 \oplus \cdots \oplus Z_r \oplus F_1 \oplus \cdots \oplus F_u,$$

where each Z_i is a non-trivial cyclic module of prime-power order, and each F_i is a non-trivial torsion-free cyclic module.

DECOMPOSITION THEOREMS

If $M = Z'_1 \oplus \cdots \oplus Z'_s \oplus F'_1 \oplus \cdots \oplus F'_v$ is another such decomposition, then $r = s$, $u = v$ and the summands Z'_j can be renumbered so that $o(Z_i) = o(Z'_i)$ for $1 \leqslant i \leqslant r$.

Proof. The existence statement is an immediate consequence of 8.2 and 8.11. The former tells us that M is a direct sum of cyclic modules, and all we have to do is to use 8.11 to break up the torsion modules among these into direct sums of cyclic modules of prime-power order.

Next we prove the uniqueness statement – there is nothing essentially new here either. By a now familiar argument we have

$$T = Z_1 \oplus \cdots \oplus Z_r = Z'_1 \oplus \cdots \oplus Z'_s$$

is the torsion submodule of M, and $u = v$, since these are both equal to the rank of M/T.

Let $\{p_1, \ldots, p_l\}$ be a set of pairwise non-associated primes such that each Z_i and Z'_j has order a power of some p_k. Renumber the Z_i so that $Z_1, \ldots, Z_{i_1}$ are of p_1-power order, $Z_{i_1+1}, \ldots, Z_{i_2}$ are of p_2-power order and so on, and renumber the Z'_j as $Z'_1, \ldots, Z'_{j_1}$, $Z'_{j_1+1}, \ldots, Z'_{j_2}, \ldots$ similarly. By collecting together the summands in the two decompositions in this way we obtain two primary decompositions of T. It follows from 8.10 that

$$Z_{i_t+1} \oplus \cdots \oplus Z_{i_{t+1}} = Z'_{j_t+1} \oplus \cdots \oplus Z'_{j_{t+1}} \tag{9}$$

this being the p_{t+1}-component of T, for $0 \leqslant t < l$. (We have put $i_0 = j_0 = 0$.) Since each summand has order a power of p_t, we can arrange the summands in each decomposition so that the order of each divides the order of its successor. Then by 8.5 we find that the number of summands on each side of (9) is the same, and that, if the modules are paired up in the obvious way, their order ideals are equal. This proves the result.

We conclude this chapter by showing that the decomposition of 8.14 is 'atomic' – the summands can be broken up no further.

8.15. Definition. An R-module M is called *indecomposable* if $M \neq \{0\}$ and M has no non-trivial direct decomposition, that is whenever $M = M_1 \oplus M_2$ is a direct sum of submodules M_1 and M_2, then either $M_1 = \{0\}$ or $M_2 = \{0\}$.

8.16. Theorem. *Every non-trivial cyclic R-module of prime power order and every non-trivial torsion-free cyclic R-module is indecomposable.*

To prove this we need the following lemma about the structure of cyclic modules of prime-power order:

8.17. Lemma. *Let $Z = Rz$ be a cyclic R-module of prime-power order p^α. Then the only submodules of Z are*

$$\{0\} = Z_\alpha \subset Z_{\alpha-1} \subset \cdots \subset Z_1 \subset Z_0 = Z,$$

where $Z_\beta = p^\beta Z$.

Proof. By 6.11 $Z \cong {}_R(R/p^\alpha R)$. In this isomorphism any submodule of Z corresponds (by 5.13) to a submodule of ${}_R R$ containing $p^\alpha R$. Such a submodule is an ideal of R, and so has the form rR where, since $rR \supseteq p^\alpha R$, $r | p^\alpha$. Therefore by the unique factorization theorem, $r = up^\beta$ where $0 \leq \beta \leq \alpha$, and u is a unit which we can arrange to be 1 by choosing r suitably. This shows that the Z_β are the only submodules of Z. Since Z_β has order exactly $p^{\alpha-\beta}$ if $0 \leq \beta \leq \alpha$, the submodules listed are all distinct.

Proof of Theorem 8.16. (i) Let Z be cyclic of prime-power order p^α with $\alpha > 0$, and suppose $Z = Z' \oplus Z''$. If Z' and Z'' are both non-zero, then an examination of the list of submodules of Z given above shows that they both contain the non-zero submodule $Z_{\alpha-1}$ and so intersect non-trivially. Therefore one of Z' and Z'' must be zero.

(ii) Since every non-trivial torsion-free cyclic R-module is isomorphic to ${}_R R$, it will suffice to show that ${}_R R$ is indecomposable. If, however, ${}_R R = R_1 \oplus R_2$, where each R_i is a non-zero submodule of R, then choosing $0 \neq r_i \in R_i$ for $i = 1, 2$ we find that $r_1 r_2 \in RR_2 \subseteq R_2$ and $r_1 r_2 = r_2 r_1 \in RR_1 \subseteq R_1$. Since R has no divisors of zero, $r_1 r_2$ is a non-trivial element of $R_1 \cap R_2$, which is a contradiction. Therefore ${}_R R$ is indecomposable.

Exercises for Chapter 8

1. Decompose the **Z**-module $\mathbf{Z}_{20} \oplus \mathbf{Z}_{40} \oplus \mathbf{Z}_{108}$ (i) into its primary components, and (ii) into indecomposable components. Try some other similar examples.

DECOMPOSITION THEOREMS

2. Find the torsion-free rank and a sequence of torsion invariants for each of the following modules:

 (i) An n-dimensional vector space V over a field $\mathbf{k}$ considered as a $\mathbf{k}$-module.
 (ii) The same space considered as a $k[x]$-module via α, where α is defined on a basis $\{v_1,\ldots,v_n\}$ of V by $\alpha v_i = v_{i+1}$ for $1 \leqslant i \leqslant n-1$ and $\alpha v_n = 0$.
 (iii) $\mathbf{Z}_p$ considered as a $\mathbf{Z}_p$-module.
 (iv) $\mathbf{Z}_p$ considered as a $\mathbf{Z}$-module.

3. Let M be a cyclic torsion module over PID. Describe the submodules of M, proving that there are only finitely many of them. Show that every quotient module of M is isomorphic to a submodule of M.

4. Use Theorems 8.2 and 8.5 to show $_R R$ is indecomposable.

5. Let L, M, N be FG p-torsion modules over a PID. Show (by considering the torsion invariants) that if $L \oplus N \cong M \oplus N$, then $L \cong M$. Extend to the case of arbitrary FG modules L, M, N. (Hint: first extend to the case when L and M are arbitrary and N is a p-torsion module.)

6*. Prove that every submodule of an FG module over PID is FG.

7. Let $M = M_1 \oplus \cdots \oplus M_t$ be a direct sum of non-trivial cyclic submodules of orders $p^{n_1},\ldots, p^{n_t}$ (where p is a prime) with $n_1 \leqslant n_2 \leqslant \cdots \leqslant n_t$. Show that the set $M(p)$ of all elements $m \in M$ satisfying $pm = 0$ is a submodule of M and may be regarded as a vector space of dimension t over the field R/pR. Suppose that $M = M_1' \oplus \cdots \oplus M_s'$ is another decomposition of M, where M_j' is cyclic of order $p^{n_j'}$ and $n_1' \leqslant \cdots \leqslant n_s'$. Show that $s = t$. By considering direct decompositions of $M/M(p)$ and using induction, prove that $n_1 = n_1',\ldots, n_t = n_t'$. This gives another proof of 8.5 for p-torsion modules. Extend to the general case.

8. Let M be an FG torsion module with invariant factor sequence $d_1, \ldots, d_s$. Show, by using 8.8 or otherwise, that no subset containing less than s elements can generate M.

In the logical framework of this book Chapter 8 rests on Chapter 7; however, the next sequence of exercises shows that the main theorems of Chapter 7 can in fact be deduced from results in Chapter 8.

9*. By making use of the splitting theorem 7.7, show that in 8.8 the assumption that M is a torsion module is unnecessary.

10. Using the previous exercise, deduce Theorem 7.1 from Theorem 8.2.

11*. Let N be a free R-module of finite rank t, and suppose $\{n_1, \ldots, n_l\}$ is a subset of N which generates N (not necessarily freely). Using the splitting theorem 7.7, show that $l \geqslant t$. Show that there exists an invertible $l \times l$ matrix $X = (x_{ij})$ such that

$$\sum_{j=1}^{l} x_{ji} n_j = n_i^* \quad (i = 1, \ldots, t)$$

and

$$\sum_{j=1}^{l} x_{ji} n_j = 0 \quad (i = t+1, \ldots, l),$$

where $\{n_1^*, \ldots, n_t^*\}$ is a basis of N.

Deduce that, if A is an $s \times t$ matrix over R, then there exists an invertible $t \times t$ matrix T over R such that $AT = (B|0)$, where the columns of B are linearly independent.

Now show that 7.10 follows from 7.1.

12. Let N be a free R-module, let $\mathbf{n} = \{n_1, \ldots, n_l\}$ be a set which generates N (not necessarily freely) and let $X = (x_{ij})$ be an invertible $l \times l$ matrix over R. Show that, if

$$n_i^* = \sum_{j=1}^{l} x_{ji} n_j \quad (i = 1, \ldots, l),$$

then $\{n_1^*, \ldots, n_l^*\}$ generates N. Deduce that 7.15 follows from 8.5.

CHAPTER NINE

Decomposition theorems – a matrix-free approach

In this chapter we shall give a proof of the fundamental theorems 8.2, 8.5 and 8.14 directly in terms of the modules themselves and not via free modules and matrices. This approach gives less computational information than the previous one, but is rather more elegant. Since no new results are proved in this chapter, it may be omitted on a first reading by the reader who is anxious to come to grips with the applications of the theory in Part III. A few results from Chapters 7 and 8 are needed, but these are all self-contained and can be read independently of the bulk of the material in those chapters.

1. Existence of the decompositions

We begin by studying the case of an FG p-torsion module over a PID R (where p is a prime in R) and proving that such a module is a direct sum of cyclic submodules. The divisibility condition of Theorem 8.2 is superfluous here since any collection of powers of a given prime p may be arranged in order so that each one divides its successor. We then consider the case of a torsion-free module, and deduce the general case from these two with very little trouble.

9.1. Lemma. *Let p be a prime in R, and let $M = Rx_1 \oplus \cdots \oplus Rx_t$ be a direct sum of cyclic modules Rx_i of order p^{α_i}, where $\alpha_1 \leqslant \cdots \leqslant \alpha_t$. Let $m \in M$, let γ be an integer with $0 \leqslant \gamma \leqslant \alpha_1$ and suppose $p^{\alpha_1-\gamma}m = 0$. Then $m = p^\gamma n$ for some $n \in M$.*

Proof. We have $m = \sum r_i x_i$ with $r_i \in R$ and so $0 = p^{\alpha_1-\gamma}m = \sum p^{\alpha_1-\gamma} r_i x_i$. It follows that $p^{\alpha_1-\gamma} r_i x_i = 0$ for each i, and so $p^{\alpha_i} | p^{\alpha_1-\gamma} r_i$. A fortiori $p^{\alpha_1} | p^{\alpha_1-\gamma} r_i$. It follows from the unique factorization theorem that a factorization of r_i into primes must contain at least γ factors associated to p, whence $p^\gamma | r_i$. Let $r_i = p^\gamma s_i$. Then $m = p^\gamma(\sum s_i x_i)$, as required.

9.2. Lemma. *Every FG p-torsion module M is a direct sum of cyclic submodules.*

Proof. It is convenient to prove instead the following stronger statement:

Let M be a FG p-torsion module generated by elements $m_1, \ldots, m_s$, where $s \geqslant 0$, m_i has order p^{α_i} and $\alpha_1 \leqslant \cdots \leqslant \alpha_s$. Then there exist elements $n_1, \ldots, n_s \in M$ such that n_i has order p^{β_i}, with $\beta_i \leqslant \alpha_i$, and $M = Rn_1 \oplus \cdots \oplus Rn_s$.

We prove this statement by induction on $\sum_{i=1}^{s} \alpha_i$, calling this the *height* of the generating set. If $\sum_{i=1}^{s} \alpha_i = 0$, then $M = \{0\}$ and the result is trivial.

We may therefore assume $\sum_{i=1}^{s} \alpha_i > 0$ and that the theorem holds for modules having a generating set of smaller height. By omitting trivial summands we may assume $\alpha_1 > 0$, and, since the theorem is obviously true if $s = 0$ or 1, we may assume that $s > 1$. Let $M^* = \sum_{i=2}^{s} Rm_i$. Then

$$M = Rm_1 + M^*. \qquad (1)$$

Since $\alpha_1 > 0$, the height of the given generating set for M^* is less than $\sum_{i=1}^{s} \alpha_i$. Hence by our inductive hypothesis there exist $n_2, \ldots, n_s \in M^*$ such that $M^* = Rn_2 \oplus \cdots \oplus Rn_s$ and n_i has order p^{β_i} with $\beta_i \leqslant \alpha_i$. Some of the n_i may of course be zero. Now $\{m_1, n_2, \ldots, n_s\}$ generates M by (1) and has height $\alpha_1 + \sum_{i=2}^{s} \beta_i$. If $\beta_i < \alpha_i$ for some i, this is less than $\sum_{i=1}^{s} \alpha_i$, and our inductive hypothesis then gives the result we want.

Therefore we may suppose that $\beta_i = \alpha_i$ for $i = 2, \ldots, s$. Now the element $m_1 + M^*$ of M/M^* satisfies $p^{\alpha_1}(m_1 + M^*) = M^*$ and so has order ideal rR, where $r | p^{\alpha_1}$. Since p is prime, $m + M^*$ therefore has an order p^γ, where $0 \leqslant \gamma \leqslant \alpha_1$. This means that

$$xm_1 \in M^* \Leftrightarrow p^\gamma | x. \qquad (2)$$

A MATRIX-FREE APPROACH

In particular, $p^\gamma m_1 = m^* \in M^*$. Now $0 = p^{\alpha_1} m_1 = p^{\alpha_1 - \gamma} m^*$. Since $\beta_2 = \alpha_2 \geqslant \alpha_1$, we have *a fortiori* $p^{\beta_2 - \gamma} m^* = 0$. Now apply 9.1 to M^*. It follows that $m^* = p^\gamma \bar{m}$ for some $\bar{m} \in M^*$. Therefore, setting $n_1 = m_1 - \bar{m}$, we have $p^\gamma n_1 = 0$. We claim that

$$M = Rn_1 \oplus M^*. \tag{3}$$

This will then complete the proof; for $M^* = Rn_2 \oplus \cdots \oplus Rn_s$, n_i has order $p^{\beta_i} = p^{\alpha_i}$ for $i = 2, \ldots, s$, and n_1 has order dividing p^γ with $\gamma \leqslant \alpha_1$. (In fact, the order of n_i is precisely p^γ.) Now $Rn_1 + M^*$ contains $n_1 + \bar{m} = m_1$ and so is the whole of M by (1). On the other hand, suppose $yn_1 \in Rn_1 \cap M^*$, with $y \in R$. Then $ym_1 = yn_1 + y\bar{m} \in M^*$. Therefore by (2) $p^\gamma | y$. Hence $yn_1 = 0$, and so $Rn_1 \cap M^* = \{0\}$. This establishes (3) and completes the proof.

We next attack the case of torsion-free modules.

9.3. Lemma. *Every FG torsion-free R-module is free of finite rank.*

Proof. Let M be such a module generated by $\{m_1, \ldots, m_s\}$. If $s = 0$, the result is clear, so we may suppose $s > 0$. We first claim that no linearly independent subset of M can contain more than s elements. For by 6.10 there exists an epimorphism ϵ of a free module E of rank s on to M. Any set of $s + 1$ elements in M has the form $\{\epsilon(e_1), \ldots, \epsilon(e_{s+1})\}$ with $e_i \in E$. By the proof of 7.2 $\{e_1, \ldots, e_{s+1}\}$ is linearly dependent, and so $\sum_{i=1}^{s+1} x_i e_i = 0$ for some $x_i \in R$, not all zero. Therefore $\sum_{i=1}^{s+1} x_i \epsilon(e_i) = 0$, and every set of $s + 1$ elements in M is linearly dependent. We may therefore choose a linearly independent set $\{f_1, \ldots, f_l\}$ in M with l as large as possible. The submodule F of M generated by these elements is free by 6.8. Now for each i the set $\{f_1, \ldots, f_l, m_i\}$ is linearly dependent, and so we have a relation

$$\sum_{j=1}^{l} r_{ji} f_j + r_i m_i = 0$$

in which not all the coefficients are zero. This means, since $\{f_1, \ldots, f_l\}$ is linearly independent, that $r_i \neq 0$ and that $r_i m_i \in F$. Let $r = r_1 \ldots r_s$. Then $r \neq 0$ and $rm_i \in F$ for all i. Hence $rm \in F$ for all $m \in M$. The map $m \to rm$ is a module endomorphism of M and maps M onto some submodule of F. Furthermore, since M is torsion-free, the kernel of this endomorphism is $\{0\}$. Therefore M is isomorphic to a submodule of F, and by 7.8 this submodule is free.

We are now ready to prove Theorem 8.2, which the reader may find it useful to refer back to on p. 124.

Proof of Theorem 8.2. Let T be the torsion submodule of M. Then by 6.1 and 6.5 M/T is an FG torsion-free module and so is free of finite rank by 9.3. Hence, by the splitting property of free modules (7.7), we have
$$M = T \oplus F \tag{4}$$
for some free submodule F of finite rank.

Let us now consider T. $T \cong M/F$ and so is FG by 6.1. Let T be generated by $\{x_1, \ldots, x_l\}$. Each x_i is a torsion element, and so $r_i x_i = 0$ for some $0 \neq r_i \in R$. Let $r = r_1 \ldots r_l$. Then $r \neq 0$. Any element of T has the form $\sum s_i x_i$ with $s_i \in R$. But then $r(\sum s_i x_i) = \sum s_i r x_i = 0$, and therefore $rT = \{0\}$. If r is a unit, then already we have $T = \{0\}$. Otherwise, by the unique factorization theorem we have
$$r = u p_1^{\alpha_1} \ldots p_k^{\alpha_k},$$
where u is a unit and the p_i are pairwise non-associated primes in R. By 8.10 we have
$$T = T_1 \oplus \cdots \oplus T_k,$$
where T_i is an FG p_i-torsion module. Therefore by Lemma 9.2 we have
$$\begin{aligned} T_1 &= T_{11} + \cdots + T_{1n}, \\ T_2 &= T_{21} + \cdots + T_{2n}, \\ &\quad \cdot \quad \cdot \quad \cdot \quad \cdot \\ T_k &= T_{k1} + \cdots + T_{kn}, \end{aligned}$$
where T_{ij} is a cyclic module of order $p_i^{\alpha_{ij}}$ and
$$\alpha_{i1} \leqslant \alpha_{i2} \leqslant \cdots \leqslant \alpha_{in}. \tag{5}$$

Here some of the T_{ij} may be zero – it is convenient to add on at the front some zero modules if necessary to get the same number of summands in each T_i.

A MATRIX-FREE APPROACH

Let $M_j = T_{1j} \oplus \cdots \oplus T_{kj}$, the sum of the modules in the j-th column. Then by 8.13 M_j is a cyclic module of order $d_j = p_1^{\alpha_{1j}} p_2^{\alpha_{2j}} \cdots p_k^{\alpha_{kj}}$. Thus from (5) we get $d_1 | d_2 | \ldots | d_n$. Also $T = M_1 \oplus \cdots \oplus M_n$. Now our free submodule F is the direct sum $M_{n+1} \oplus \cdots \oplus M_s$ of non-trivial torsion-free cyclic modules which have order 0. Hence by (4)

$$M = M_1 \oplus \cdots \oplus M_s$$

which is precisely the required decomposition of 8.2 with $d_{n+1} = \cdots = d_s = 0$.

Notice that we have also proved the existence statement of 8.14. For M is the direct sum of the T_{ij}, which are cyclic of prime-power order, and $M_{n+1}, \ldots, M_s$, which are torsion-free cyclic.

2. Uniqueness – a cancellation property of FG modules

We wish to prove the main uniqueness statements of Chapter 8 (8.5 and the second part of 8.14). Essentially these say that in either type of decomposition the summands are unique 'to within isomorphism'. The argument which we shall employ will be one which uses induction on the number of summands together with the 'cancellation lemma', which essentially reduces the problem to that of showing that in two direct decompositions of the type considered there is a summand in the first decomposition which is isomorphic to one in the second.

9.4. Lemma. *Let T_1, T_2 be isomorphic FG torsion modules (over a PID R as usual), let N_1, N_2 be any R-modules, and suppose that*

$$T_1 \oplus N_1 \cong T_2 \oplus N_2.$$

Then $N_1 \cong N_2$.

In other words, under suitable circumstances we can cancel isomorphic summands from direct decompositions. The proof of this result is based on some elementary facts about cyclic modules of prime-power order.

9.5. Lemma. *Let Z be a cyclic R-module of prime-power order and let ϕ, ψ be endomorphisms of Z such that $\phi + \psi = 1$, the identity endomorphism of Z. Then one of ϕ and ψ is an automorphism of Z.*

Proof. Let Z have order p^α. We may clearly suppose $Z \neq \{0\}$ and so $\alpha > 0$. Then by 8.17 Z has a unique non-zero submodule $Z_{\alpha-1} = p^{\alpha-1}Z$ which is contained in every non-zero submodule of Z. Therefore, if $\ker \phi$ and $\ker \psi$ were both non-zero, they would both contain $Z_{\alpha-1}$, which would then be mapped to zero by $\phi + \psi$. But this is impossible, since $\phi + \psi = 1$. Therefore one (at least) of ϕ and ψ has kernel $\{0\}$.

If $\ker \phi = \{0\}$, then $\operatorname{im} \phi$ is a submodule of Z isomorphic to Z itself. Again by 8.17 such a submodule has the form $p^\beta Z$ with $0 \leq \beta \leq \alpha$. This submodule clearly has order $p^{\alpha-\beta}$, and so, since isomorphic cyclic modules have the same order ideal, we must have $p^{\alpha-\beta} \sim p^\alpha$ and therefore $\beta = 0$. Hence $\operatorname{im} \phi = Z$ and ϕ is an automorphism.

We also need some elementary facts about direct decompositions. Let

$$M = M_1 \oplus M_2 \tag{6}$$

be a module expressed as a direct sum of two submodules M_1 and M_2. We recall that the projections π_i of M on M_i associated with this decomposition are defined by $\pi_i(m) = m_i$, where $m = m_1 + m_2$ with $m_i \in M_i$. Since such expressions are unique, π_i is well-defined and is easily seen to be an endomorphism of M with kernel M_j ($j \neq i$). The reader will have no trouble in verifying the following statements:

(i) $\pi_1 + \pi_2 = 1$, the identity endomorphism of M.

(ii) $\pi_i \pi_j = 0$ if $i \neq j$ (where of course 0 is the endomorphism of M which maps every element to zero).

(iii) $\pi_i^2 = \pi_i$ for $i = 1, 2$.

9.6. Lemma. *If M is as in (6), and N is a submodule of M containing M_1, then $N = M_1 \oplus (N \cap M_2)$.*

Proof. Let $n \in N$. Then we can write $n = m_1 + m_2$ with $m_i \in M_i$, and, since $M_1 \subseteq N$, it follows that $m_2 = n - m_1 \in N$. Therefore $n \in M_1 + (N \cap M_2)$. Since the two summands are both contained in N and their intersection is $\{0\}$, the result follows.

We are now ready to prove Lemma 9.4.

A MATRIX-FREE APPROACH

Proof of Lemma 9.4. We first notice that it is sufficient to consider the case when T_1 and T_2 are actually cyclic modules of prime-power order. For in the general case $T_1 = Z_{11} \oplus \cdots \oplus Z_{1t}$ is a direct sum of cyclic modules of prime-power order by the existence statement of 8.14. If ϵ denotes an isomorphism from T_1 to T_2, we have $T_2 = Z_{21} \oplus \cdots \oplus Z_{2t}$, where $Z_{2j} = \epsilon(Z_{1j}) \cong Z_{1j}$. Then

$$Z_{11} \oplus \cdots \oplus Z_{1t} \oplus N_1 \cong Z_{21} \oplus \cdots \oplus Z_{2t} \oplus N_2,$$

and, if we know that isomorphic cyclic modules of prime-power order can be cancelled, then we can cancel the pairs Z_{1j}, Z_{2j} successively to conclude that $N_1 \cong N_2$.

We therefore assume

$$Z_1 \oplus N_1 \cong Z_2 \oplus N_2, \tag{7}$$

where Z_1 and Z_2 are isomorphic cyclic modules of prime-power order, and have to show $N_1 \cong N_2$. We may clearly suppose $Z_1 \neq \{0\}$. Let $\theta : Z_1 \oplus N_1 \to Z_2 + N_2$ be the isomorphism of (7). Then $Z_2 \oplus N_2 = \theta(Z_1 \oplus N_1) = \theta(Z_1) \oplus \theta(N_1)$, and $\theta(Z_1) \cong Z_1 \cong Z_2$, $\theta(N_1) \cong N_1$. Thus it is enough to show that $\theta(N_1) \cong N_2$, and by replacing Z_1, N_1 by $\theta(Z_1)$, $\theta(N_1)$ we may therefore assume

$$Z_1 \oplus N_1 = Z_2 \oplus N_2, = M \text{ say}.$$

Let ζ_1, ν_1 be the projections of M on Z_1 and N_1 respectively associated with the first decomposition, and let ζ_2, ν_2 be defined similarly with respect to the second decomposition. Consider the endomorphism $\zeta_1(\zeta_2 + \nu_2) = \zeta_1 \zeta_2 + \zeta_1 \nu_2$. It is equal to ζ_1, since $\zeta_2 + \nu_2 = 1$, and therefore its restriction to Z_1 is obviously the identity automorphism of Z_1. Hence by 9.5 one of $\zeta_1 \zeta_2$ and $\zeta_1 \nu_2$, when restricted to Z_1, induces an automorphism on Z_1.

Case 1. $\zeta_1 \zeta_2|_{Z_1}$ is an automorphism of Z_1 (this notation denotes the restriction of $\zeta_1 \zeta_2$ to Z_1). Let $Z_2' = \zeta_2(Z_1)$. We claim that

$$M = Z_2' \oplus N_1. \tag{8}$$

Now in the first place any element of Z_2' has the form $\zeta_2(z_1)$ with $z_1 \in Z_1$. If such an element belongs also to $N_1 = \ker \zeta_1$, then $0 = \zeta_1 \zeta_2(z_1)$. But $\zeta_1 \zeta_2$ is an automorphism of Z_1 so that $z_1 = 0$. Hence $\zeta_2(z_1) = 0$. This shows that $Z_2' \cap N_1 = \{0\}$.

To see that $Z_2' + N_1 = M$ we notice that, since we certainly have $Z_1 + N_1 = M$, any element $m \in M$ has the form $m = \bar{z}_1 + n_1$ with

$\bar{z}_1 \in Z_1$ and $n_1 \in N_1$. Now $\zeta_1 \zeta_2$ maps Z_1 onto itself; therefore ζ_1 must map $Z'_2 = \zeta_2(Z_1)$ onto the whole of Z_1. Hence $\bar{z}_1 = \zeta_1(z'_2)$ for some $z'_2 \in Z'_2$. Therefore $\zeta_1(m) = \bar{z}_1 = \zeta_1(z'_2)$. Hence $\zeta_1(m - z'_2) = 0$ and $m - z'_2 \in \ker \zeta_1 = N_1$. Therefore $m \in Z'_2 + N_1$, as required. We have now established (8). Now $Z'_2 \subseteq \operatorname{im} \zeta_2 \subseteq Z_2$, and so $Z_2 = Z'_2 \oplus (Z_2 \cap N_1)$ by 9.6. But Z_2 is indecomposable by 8.16; since $Z'_2 \neq 0$, we therefore have $Z_2 \cap N_1 = \{0\}$, $Z_2 = Z'_2$ and $M = Z_2 \oplus N_1$ from (8). Therefore $M/Z_2 = N_1 \oplus Z_2/Z_2 \cong N_1$ by 5.11. Since $M = Z_2 \oplus N_2$, we similarly have $M/Z_2 \cong N_2$. Therefore $N_1 \cong N_2$ in this case.

Case 2. $\zeta_1 \nu_2|_{Z_1}$ is an automorphism. In this case let $Z''_2 = \nu_2(Z_1)$. Then an argument similar to the above with ζ_2 replaced by ν_2 gives
$$M = Z''_2 \oplus N_1. \tag{9}$$
This time we have $Z''_2 \subseteq N_2$, and so by 9.6,
$$N_2 = Z''_2 \oplus N_3, \tag{10}$$
where $N_3 = N_1 \cap N_2$. Therefore
$$M = Z_2 \oplus N_2 = Z_2 \oplus Z''_2 \oplus N_3. \tag{11}$$

Now from (9) $N_1 \cong M/Z''_2$. From (11) $M/Z''_2 \cong Z_2 \oplus N_3$. But (9) gives $Z''_2 \cong M/N_1 \cong Z_1$, and by assumption $Z_1 \cong Z_2$. Hence $Z_2 \oplus N_3 \cong Z''_2 \oplus N_3 = N_2$ from (10). The final outcome of this chain of isomorphisms is that $N_1 \cong N_2$, which is what we wanted to prove. The proof is therefore complete.

Remark. The argument which we have given above lies at the root of many theorems about uniqueness of direct decompositions in other contexts. We shall give an example in Exercise 1 at the end of the chapter to show that 9.4 does not hold without restriction on the modules T_i.

Proof of the uniqueness theorems. These are 8.5 and the uniqueness part of 8.14. We consider first 8.5. Here we have
$$M = M_1 \oplus \cdots \oplus M_s = M'_1 \oplus \cdots \oplus M'_t,$$
where M_i, M'_j are non-trivial cyclic modules of orders d_i, d'_j respectively and $d_1|d_2|\cdots|d_s$, $d'_1|d'_2|\cdots|d'_t$. Our proof will be by

A MATRIX-FREE APPROACH

induction on s, the case $s=0$ being clear. We may therefore assume $s>0$; then also $t>0$. Let $u+1$ be the first integer i such that $d_i = 0$, and let v be similarly defined with respect to the second decomposition. Then by the argument of 8.3

$$T = M_1 \oplus \cdots \oplus M_u = M_1' \oplus \cdots \oplus M_v' \qquad (12)$$

is the torsion submodule of M. Thus $M/T \cong M_{u+1} \oplus \cdots \oplus M_s$ is free of rank $s-u$, and similarly M/T is free of rank $t-v$. Therefore

$$s - u = t - v \qquad (13)$$

by 7.2. Now if $u < s$, then by induction we obtain $u = v$ and $d_1 \sim d_1', \ldots, d_u \sim d_u'$. (13) then gives $s = t$, and since $d_{u+1}, \ldots, d_s$, $d_{v+1}', \ldots, d_t'$ are all zero, the proof is complete in this case.

Hence we may assume $u = s$; then also from (13) $v = t$, and M is a torsion module. Now $d_s M = \sum_{i=1}^{s} d_s M_i = \{0\}$, since the order d_i of M_i divides d_s. Therefore certainly $d_s M_t' = \{0\}$, and it follows that $d_t' | d_s$. Similarly $d_s | d_t'$, and so $d_s \sim d_t'$. Therefore the order ideals $d_s R$ and $d_t' R$ of M_s and M_t' are equal, and so $M_s \cong M_t'$ by 6.11. Hence by 9.4 we have $M_1 \oplus \cdots \oplus M_{s-1} \cong M_1' \oplus \cdots \oplus M_{t-1}'$. Since we may replace isomorphism here by equality in the usual way, our inductive hypothesis tells us that $s-1 = t-1$, that is $s = t$, and $d_1 \sim d_1', \ldots, d_{s-1} \sim d_{s-1}'$. Since $d_s \sim d_s'$, the theorem is proved.

The uniqueness statement of 8.14 can now be deduced by the argument given to prove it in the last chapter (see p. 135).

Exercises for Chapter 9

1. Let M be the set of all infinite sequences $(z_1, z_2, \ldots)$ with $z_i \in \mathbf{Z}$, regarded as a $\mathbf{Z}$-module by defining

$$(z_1, z_2, \ldots) + (z_1', z_2', \ldots) = (z_1 + z_1', z_2 + z_2', \ldots)$$

$$z(z_1, z_2, \ldots) = (zz_1, zz_2, \ldots).$$

Show that $\mathbf{Z} \oplus M \cong M \cong \{0\} \oplus M$ and deduce that 9.4 does not hold without restriction on the modules T_i. (M is the external direct sum of $\mathbf{Z}$ with itself a countable infinity of times). Show that nevertheless 9.4 does hold for arbitrary *finitely-generated* modules T_i (over a PID).

2. Let M be an FG p-torsion module (over a PID R as usual). Suppose that $p^\alpha M = \{0\}$ and that x is an element of M of order exactly p^α. Show that $M = N \oplus Rx$ for some submodule N.

(*Hint*: Let $M = Rx_1 \oplus \cdots \oplus Rx_t$ as in 8.2, where x_i has order p^{α_i} and $\alpha_1 \leqslant \alpha_2 \leqslant \cdots \leqslant \alpha_t$. Show that $\alpha = \alpha_t$. Write $x = \sum r_i x_i$ ($r_i \in R$), and show that $(r_i, p) = [1]$ for some i such that $\alpha_i = \alpha$. Take $N = \sum_{j \neq i} Rx_j$.)

3**. Answer Exercise 2 above without using 8.2. (Consider first the case when M is generated by x and one other element, and then use induction on the number of generators.) Deduce Lemma 9.2.

PART THREE

Applications to groups and matrices

CHAPTER TEN

Finitely-generated Abelian groups

1. Z-modules

We described in §1 of Chapter 5 how the structure of a **Z**-module could be imposed on an arbitrary Abelian group A. If $0 < n \in \mathbf{Z}$ and $a \in A$, the **Z**-action is given by

$$0a = 0,$$
$$\left.\begin{array}{l}na = (a + \cdots + a),\\ (-n)a = -(a + \cdots + a)\end{array}\right\} \text{ with } n \text{ terms } a.$$

This poses the converse question: is every **Z**-module M just an Abelian group with the **Z**-action just defined? The module axioms quickly show that this is the case. By Remark 3 on p. 71 we have $0m = 0$ for all $m \in M$. Moreover, if $0 < n \in \mathbf{Z}$, by axioms **M2** and **M4** we have

$$nm = (1 + \cdots + 1)m = m + \cdots + m, \quad \text{with } n \text{ terms } m.$$

Again by Remark 3 we have

$$(-n)m = -(nm) = -(m + \cdots + m), \quad \text{with } n \text{ terms } m.$$

This is precisely the **Z**-action defined at the beginning of the paragraph. Therefore **Z**-modules are no more nor less than Abelian groups – old bottles with new labels.

Let B be a subgroup of an Abelian group A. If A is considered as a **Z**-module, then as we saw in Chapter 5, §2, B is a submodule of A. Conversely of course, every submodule of the **Z**-module A is a subgroup of the Abelian group obtained from A by disregarding the **Z**-action.

If X is a subset of an Abelian group A, then the subgroup generated by X is the set of elements of the form $\sum n_i x_i$ with $n_i \in \mathbf{Z}$ and $x_i \in X$, and these comprise exactly the submodule $\mathbf{Z}X$ of A generated by X. Hence a finitely-generated Abelian group is precisely the same as a finitely-generated $\mathbf{Z}$-module.

We summarize these and similar elementary facts in the following conversion table for the terminology applying to these two different views of the same situation:

$\mathbf{Z}$-module	Abelian group
submodule	subgroup
quotient module	quotient group
$\mathbf{Z}$-homomorphism	group homomorphism
finitely-generated (sub-)module	finitely-generated (sub-)groups
cyclic (sub-)module	cyclic (sub-)group
element with order ideal $n\mathbf{Z}$ ($n \neq 0$)	element of order $|n|$
element with order ideal $\{0\}$	element of infinite order
free $\mathbf{Z}$-module of rank s	free Abelian group of rank s

The last entry can be viewed as a definition by those who have not already met free Abelian groups.

2. Classification of finitely-generated Abelian groups

The decomposition theorems of Part II enable us to give a full classification of finitely-generated Abelian groups. By this we mean that it is possible to associate with each such group a set of invariants which uniquely determine that group (up to isomorphism of course), and that one can write down a complete list of what the possible sets of invariants are. This list is then effectively a list of all FG Abelian groups up to isomorphism – every FG Abelian group corresponds to a set of invariants on the list and vice-versa, and two groups are isomorphic if and only if they have the same set of invariants. Moreover, one can usually compute the invariants of a given group fairly readily – for example, if a FG Abelian group is given by generators and relations, there is a systematic method whereby its invariants may be determined in a finite number of

FINITELY-GENERATED ABELIAN GROUPS

steps. For finite Abelian groups the classification will tell us exactly how many non-isomorphic groups of a given order exist.

We now specialize the decomposition theorems 8.2, 8.5 and 8.14 to the case of FG **Z**-modules, bearing in mind that **Z** is a PID, and translate the results into the group-theoretic language.

10.1. Theorem. *Let A be a finitely-generated Abelian group. Then A has a direct decomposition*

$$A = A_1 \oplus \cdots \oplus A_r \oplus A_{r+1} \oplus \cdots \oplus A_{r+t},$$

where (i) A_i *is a non-trivial finite cyclic group of order n_i for $i = 1, \ldots, r$,*

(ii) A_i *is an infinite cyclic group for $i = r+1, \ldots, r+t$, and*

(iii) $n_1 | n_2 | \cdots | n_r$.

The integers t and $n_1, \ldots, n_r$ occurring in such a decomposition are uniquely determined by A.

Remarks. 1. According to 8.5 only the ideals $n_1 \mathbf{Z}, \ldots, n_r \mathbf{Z}$ are uniquely determined. But the order n_i of A_i is by definition the positive generator of its order ideal, and this is uniquely determined. Of course, we cannot speak of positive and negative elements in a general PID.

2. The number t is the torsion-free rank of A, and $n_1, \ldots, n_r$ is a sequence of torsion invariants of A. In fact, by making all the torsion invariants positive we can (and will) speak of *the* sequence of torsion invariants of A.

10.2. Corollary. *Two finitely-generated Abelian groups are isomorphic if and only if they have the same torsion-free rank and the same sequence of torsion invariants. If t and r are integers ≥ 0 and $n_1 | \cdots | n_r$ is a sequence of integers > 1, then there exists a finitely-generated Abelian group with torsion-free rank t and torsion invariants $n_1, \ldots, n_r$.*

Proof. Most of this we have already proved. To construct a group with given torsion-free rank and torsion invariants we simply form an external direct sum of t infinite cyclic groups together with cyclic groups of orders $n_1, \ldots, n_r$.

This achieves the classification mentioned – we associate with each FG Abelian group its torsion-free rank and sequence of torsion invariants. Another classification can be obtained by transposing Theorem 8.14:

10.3. Theorem. *Let A be a finitely-generated Abelian group. Then A has a direct decomposition*
$$A = B_1 \oplus \cdots \oplus B_s \oplus B_{s+1} \oplus \cdots \oplus B_{s+t},$$
where (i) B_i *is a non-trivial cyclic group of prime-power order $p_i^{\alpha_i}$ for $i = 1, \ldots, s$, and*

(ii) B_i *is an infinite cyclic group for $i = s + 1, \ldots, s + t$.*

In any such decomposition the integer t is uniquely determined and the prime-power orders $p_i^{\alpha_i}$ are determined to within rearrangement.

Notice that the primes p_i are not assumed to be all distinct, and that t is again the torsion-free rank of A.

10.4. Definition. The prime powers occurring in 10.3 are called the *primary invariants* of A.

10.5. Corollary. *Two finitely-generated Abelian groups are isomorphic if and only if they have the same torsion-free rank and the same primary invariants. There exists a finitely-generated Abelian group having as torsion-free rank a given integer $t \geqslant 0$ and as primary invariants a given finite set of prime powers > 1.*

3. Finite Abelian groups

Notation. To emphasize the group-theoretic context we shall use C_n (rather than $\mathbf{Z}_n$) to denote a cyclic group of order $n \geqslant 1$; C_0 will denote an infinite cyclic group (because, as **Z**-module, its order ideal is generated by 0).

$|A|$ denotes the order of the finite group A, that is, the number of elements in A. If A happens to be cyclic, this coincides with the order of A in the previous sense.

Notice that, if A and B are finite Abelian groups, then $|A \oplus B| = |A| \cdot |B|$, since the elements of $A \oplus B$ can be identified with ordered pairs (a, b) with $a \in A$, $b \in B$. Therefore
$$|C_{n_1} \oplus \cdots \oplus C_{n_r}| = n_1 \ldots n$$

if $r \geq 1$ and $n_i \neq 0$. Thus each Abelian group of order $n > 1$ has as torsion invariants a sequence $n_1 | n_2 | \cdots | n_r$ with $r > 0$, $n_i > 1$ and $n_1 \ldots n_r = n$, and by writing down a list of all such sequences we obtain a list (up to isomorphism) of all Abelian groups of order n.

Example. There are two Abelian groups of order 12, namely C_{12} and $C_2 \oplus C_6$, with torsion invariants 12 and 2, 6 respectively. By 8.11 we have
$$C_{12} = C_4 \oplus C_3$$
and
$$C_2 \oplus C_6 = C_2 \oplus C_2 \oplus C_3,$$
and so the primary invariants of these two groups are $\{2^2, 3\}$ and $\{2, 2, 3\}$ respectively.

It is usually easier, in fact, to begin by determining the possibilities for the primary invariants of an Abelian group of order $n > 1$. If A is such a group, then
$$A = A_1 \oplus \cdots \oplus A_l,$$
where each A_i is a non-trivial cyclic group of prime-power order. If $p_1, \ldots, p_k$ are the *distinct* (positive) primes occurring and if we renumber the summands as A_{ij}, where A_{ij} has order $p_i^{\alpha_{ij}}$ and $\alpha_{i1} \leq \alpha_{i2} \leq \ldots$, then the sum B_i of the summands of p_i-power order has $|B_i| = p_i^{\alpha_i}$, where $\alpha_i = \sum_j \alpha_{ij}$, and
$$A = B_1 \oplus \cdots \oplus B_k.$$
Therefore $n = |A| = p_1^{\alpha_1} \cdots p_k^{\alpha_k}$, and this must be the unique factorization of n into positive primes. The possibilities for the primary invariants of an Abelian group of order n are thus obtained by determining for each i the various sequences $\alpha_{i1} \leq \alpha_{i2} \leq \cdots$ such that
$$\alpha_{i1} + \alpha_{i2} + \cdots = \alpha_i$$
with each $\alpha_{ij} \geq 1$, and combining these together in all possible ways. A numerical illustration will make this clear.

Worked Example. Find all the Abelian groups of order 360 (up to isomorphism), giving the primary invariants and torsion invariants of each.

First we notice that the prime factorization of 360 is $2^3 . 3^2 . 5$. If A is an Abelian group of order 360 with primary invariants

$\{2^{\alpha_1}, 2^{\alpha_2}, \ldots, 3^{\beta_1}, 3^{\beta_2}, \ldots, 5^{\gamma_1}, \ldots\}$, then as above we have $360 = 2^{\Sigma \alpha_i} 3^{\Sigma \beta_i} 5^{\Sigma \gamma_i}$ whence

$$\alpha_1 + \alpha_2 + \cdots = 3,$$
$$\beta_1 + \beta_2 + \cdots = 2,$$
$$\gamma_1 + \gamma_2 + \cdots = 1.$$

Furthermore, we assume $\alpha_1 \leq \alpha_2 \leq \ldots$, etc., and that $\alpha_i, \beta_i, \gamma_i \geq 1$. The possibilities are then

2-exponents: $\{\alpha_1, \ldots\} = \{3\}, \{1, 2\}$ or $\{1, 1, 1\}$.
3-exponents: $\{\beta_1, \ldots\} = \{2\}$ or $\{1, 1\}$.
5-exponents: $\{\gamma_1, \ldots\} = \{1\}$.

Combining these in all possible ways, we obtain six $(= 3.2.1)$ mutually non-isomorphic candidates for A. These groups, with their primary invariants, are:

$A_1 = C_8 \oplus C_9 \oplus C_5;$ $\qquad \{2^3, 3^2, 5\}$.
$A_2 = C_8 \oplus C_3 \oplus C_3 \oplus C_5;$ $\qquad \{2^3, 3, 3, 5\}$.
$A_3 = C_2 \oplus C_4 \oplus C_9 \oplus C_5;$ $\qquad \{2, 2^2, 3^2, 5\}$.
$A_4 = C_2 \oplus C_4 \oplus C_3 \oplus C_3 \oplus C_5;$ $\qquad \{2, 2^2, 3, 3, 5\}$.
$A_5 = C_2 \oplus C_2 \oplus C_2 \oplus C_9 \oplus C_5;$ $\qquad \{2, 2, 2, 3^2, 5\}$.
$A_6 = C_2 \oplus C_2 \oplus C_2 \oplus C_3 \oplus C_3 \oplus C_5;$ $\qquad \{2, 2, 2, 3, 3, 5\}$.

This is really a shorthand notation meaning that each A_i is a direct sum of subgroups isomorphic to the C_n's written.

To obtain a decomposition giving the torsion invariants we select a summand of highest order from each primary component, and combine these to get the summand of highest order in the torsion-invariant decomposition; we then take a summand of next highest order (if there is one) from each primary component and combine these, and so on. In this way we derive (using 8.13) the following decompositions in the style of 10.1, with the torsion invariants as given:

$A_1 = C_8 \oplus C_9 \oplus C_5 = C_{360};$ $\qquad$ 360.
$A_2 = C_3 \oplus (C_8 \oplus C_3 \oplus C_5) = C_3 \oplus C_{120};$ $\qquad$ 3, 120.
$A_3 = C_2 \oplus (C_4 \oplus C_9 \oplus C_5) = C_2 \oplus C_{180};$ $\qquad$ 2, 180.
$A_4 = (C_2 \oplus C_3) \oplus (C_4 \oplus C_3 \oplus C_5) = C_6 \oplus C_{60};$ $\qquad$ 6, 60.
$A_5 = C_2 \oplus C_2 \oplus (C_2 \oplus C_9 \oplus C_5) = C_2 \oplus C_2 \oplus C_{90};$ $\qquad$ 2, 2, 90.
$A_6 = C_2 \oplus (C_2 \oplus C_3) \oplus (C_2 \oplus C_3 \oplus C_5) = C_2 \oplus C_6 \oplus C_{30};$
$\qquad$ 2, 6, 30.

FINITELY-GENERATED ABELIAN GROUPS

4. Generators and relations

If we are simply told that a given Abelian group A is generated by k elements $a_1,\ldots, a_k \in A$, then we have very little information about A – certainly far from enough to determine A up to isomorphism. For example, if $k = 1$, we just know that A is cyclic – it could be of infinite order or of any finite order.

How much more information do we need in order to describe completely an Abelian group generated by given elements $a_1,\ldots, a_k$? Now the information we have tells us that any element of A can be expressed in the form $\sum n_i a_i$ with $n_i \in \mathbf{Z}$, but does not tell us when two different expressions of this form represent the same element of A, or, in particular, whether some given expression represents 0. Of course, two expressions $\sum n_i a_i$ and $\sum n'_i a_i$ represent the same element of A if and only if their difference $\sum (n_i - n'_i)a_i$ represents 0.

We thus need to know which expressions $\sum n_i a_i$ represent the element 0, or in other words, to know which 'relations' hold between the generators. If we wrote down a complete list of all the relations, that is, of all the expressions which represent zero, then A would be completely determined – each element of A could be viewed as a 'class of expressions', two expressions belonging to the same class (or representing the same element of A) if and only if their difference is one of the listed expressions representing 0. Classes could then be added by adding their representatives.

For example, we have

$$C_6 = \langle a : 6na = 0 \text{ for all } n \in \mathbf{Z} \rangle.$$

(Read the right-hand side as 'the Abelian group generated by a subject to the relations $6na = 0$'.) For, if a generates C_6, then an expression ma represents zero if and only if $6|m$. Similarly we have

$$C_2 \oplus C_5 = \langle a, b : 2ma + 5nb = 0 \text{ for all } m, n \in \mathbf{Z} \rangle.$$

For, if an Abelian group is the direct sum of a cyclic subgroup of order 2 generated by a and a cyclic subgroup of order 5 generated by b, then an expression $ka + lb$ represents 0 if and only if $2|k$ and $5|l$.

Notice that in these two examples we could be more economical by writing

$$C_6 = \langle a : 6a = 0 \rangle, \quad \text{and} \quad C_2 \oplus C_5 = \langle a, b : 2a = 5b = 0 \rangle.$$

For, if $6a = 0$, then certainly $6na = 0$ for all integers n, and, if $2a = 5b = 0$, then $2ma + 5nb = 0$ for all integers m, n. In a case like this the expressions representing zero are to be taken as the linear combinations of the given ones (which necessarily represent zero) and no others.

This approach has two drawbacks. Firstly, what are these 'expressions'? They look as if they ought to be elements of A, but obviously they are not, since two different expressions are supposed to be able to 'represent' the same element of A. Secondly, what happens if, instead of taking a known group and determining its relations, we try to construct a group generated by a given set of elements between which given relations hold? For example, what is the meaning of

$$B = \langle a, b : 2a + 3b = a - 7b = 0 \rangle ?$$

If one tries to take this group as consisting of 'classes of expressions' $na + mb$, two expressions belonging to the same class if and only if their difference is a linear combination of $2a + 3b$ and $a - 7b$, one runs into the problem of showing that the addition of classes by representatives is well-defined.

Fortunately there is a more elegant and satisfactory approach to the whole question. Keeping in mind our idea of what the group B above should be like, let F be a free Abelian group with basis $\{x, y\}$. Then there exists an epimorphism $\epsilon : F \to B$ mapping $x \to a$ and $y \to b$. The fact that $2a + 3b = a - 7b = 0$ means that the elements $2x + 3y$ and $x - 7y$ belong to $\ker \epsilon$. The idea that only those expressions $na + mb$ which can be written as linear combinations of $2a + 3b$ and $a - 7b$ are supposed to represent zero means in precise language that $\ker \epsilon$ consists *exactly* of the linear combinations of $2x + 3y$ and $x - 7y$, in other words, that these elements *generate* $\ker \epsilon$. This puts matters on a precise footing and shows us how to define, in a precise way, what it means to present an Abelian group by generators and relations.

10.6. Definition. Let A be an Abelian group generated by s elements $a_1, \ldots, a_s$ and let $(r_{1i}, \ldots, r_{si})$ $(i = 1, \ldots, t)$ be t s-tuples of integers. We say that A has the *presentation*

$$\langle a_1, \ldots, a_s : \sum_{j=1}^{s} r_{ji} a_j = 0 \text{ for } i = 1, \ldots, t \rangle,$$

or that A is *generated by $a_1,\ldots, a_s$ subject to the defining relations* $\sum_{j=1}^{s} r_{ji} a_j = 0$ $(i = 1,\ldots,t)$ if, whenever F is a free Abelian group of rank s with a basis $\{f_1,\ldots, f_s\}$ and ϵ is the unique epimorphism $F \to A$ such that $\epsilon(f_i) = a_i$ for $i = 1,\ldots, s$, then $\ker \epsilon$ is generated by the t elements $\sum_{j=1}^{s} r_{ji} f_j$ $(i = 1,\ldots,t)$.

Remarks. 1. It is sufficient, in fact, for the above condition to be satisfied just for one basis of one free Abelian group. For assume that F has basis $\{f_1,\ldots, f_s\}$, that ϵ is the epimorphism mapping each f_i to a_i, and that $\ker \epsilon = K$ is generated by the elements $\sum_{j=1}^{s} r_{ji} f_j$ $(i = 1,\ldots,t)$. Let F' be a free Abelian group with basis $\{f'_1,\ldots, f'_s\}$, and let ϵ' be the epimorphism mapping each f'_i to a_i, with kernel K'. If ϕ is the isomorphism $F \to F'$ given by $\phi(f_i) = f'_i$ for all i, and $\psi = \phi^{-1}$, then $\epsilon = \epsilon' \phi$. Therefore $\{0\} = \epsilon(K) = \epsilon' \phi(K)$, whence $\phi(K) \subseteq K'$; similarly $\psi(K') \subseteq K$. Therefore $K' = \phi\psi(K') \subseteq \phi(K)$, and we have equality. Therefore K' is generated by the elements $\phi(\sum r_{ji} f_j) = \sum r_{ji} f'_j$.

2. It now follows that, if $(r_{1i},\ldots,r_{si})$ are t given s-tuples of integers, then there exists a group generated by s elements subject to the defining relations determined by these s-tuples. For, taking a free Abelian group F with basis $\{f_1,\ldots, f_s\}$ and letting N denote the subgroup generated by the elements $\sum r_{ji} f_j$, the quotient group F/N has the required presentation. For F/N is generated by the elements $f_i + N$, the natural homomorphism ν maps each f_i to $f_i + N$, and $\ker \nu = N$, which by construction is generated by the elements $\sum r_{ji} f_j$.

3. Notice that we have not defined the *group* generated by given elements subject to given defining relations, but the *Abelian group* so generated – the commutative law is always understood. Non-Abelian groups do not concern us in this book.

The next result shows, as we should expect, that the Abelian group generated by a given set of s elements subject to given defining relations is in a sense the 'largest' Abelian group which can be generated by s elements satisfying the relations in question. Thus, for example, the group C_3 is generated by a single element a satisfying the relation $6a = 0$, but this is not a defining relation in this case.

10.7. Lemma. *Let $A = \langle a_1,\ldots,a_s : \sum_{j=1}^{s} r_{ji} a_j = 0 \text{ for } i = 1,\ldots,t \rangle$, and let B be an Abelian group generated by $b_1,\ldots, b_s$. Suppose further* that $\sum_{j=1}^{s} r_{ji} b_j = 0$ for all i. Then there exists an epimorphism $\psi: A \to B$ such that $\psi(a_i) = b_i$ for $i = 1, \ldots, s$.

Proof. Let F be a free Abelian group with basis $\{f_1, \ldots, f_s\}$, let $\epsilon: F \to A$ be the unique epimorphism satisfying $\epsilon(f_i) = a_i$ ($1 \leq i \leq s$) and let $\phi: F \to B$ be the unique epimorphism satisfying $\phi(f_i) = b_i$ ($1 \leq i \leq s$). By 5.9 and 5.10 there exists an isomorphism $\lambda: F/\ker \epsilon \to A$ such that $\lambda \nu = \epsilon$, where ν is the natural homomorphism. Now by definition the elements $\sum r_{ji} f_j$ generate $\ker \epsilon$, and by assumption ϕ maps all these elements to zero. Therefore $\ker \phi \geq \ker \epsilon$. It follows from 5.9 that there exists a homomorphism $\mu: F/\ker \epsilon \to B$ such that $\mu \nu = \phi$.

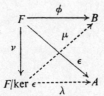

Let $\psi = \mu \lambda^{-1}$. Then $\psi(a_i) = \psi \epsilon(f_i) = \mu \lambda^{-1} \epsilon(f_i) = \mu \nu(f_i) = \phi(f_i) = b_i$, so that ψ is the required map.

5. Computing invariants from presentations

It is natural at this stage to raise the problem: given an Abelian group A in terms of a 'generators and relations' presentation, what can be said about the structure of A? For example, can we determine its torsion-free rank and torsion invariants? A given group may have many different presentations; for example, since $C_6 = C_2 \oplus C_3$,

$$\langle a : 6a = 0 \rangle \quad \text{and} \quad \langle a, b : 2a = 3b = 0 \rangle$$

are both presentations of a cyclic group of order 6. It is often important to know whether two given presentations determine isomorphic groups, and if we can recover the invariants of a group from a presentation, we shall certainly be able to decide this.

Let

$$A = \langle a_1, \ldots, a_s : \sum_{j=1}^{s} r_{ji} a_j = 0 \text{ for } i = 1, \ldots, t \rangle.$$

FINITELY-GENERATED ABELIAN GROUPS

Then $A \cong F/N$, where F is a free Abelian group with basis $\{f_1,\ldots,f_s\}$ and N is generated by the elements $\sum r_{ji} f_j$. If we can find a basis $\{f'_1,\ldots,f'_s\}$ of F such that $N = \mathbf{Z}(d_1 f'_1) \oplus \cdots \oplus \mathbf{Z}(d_s f'_s)$ for suitable integers $d_1|\cdots|d_s$ (which may be assumed non-negative), then the results of Chapter 8, §1 tell us that F/N is a direct sum of cyclic groups of orders $d_1,\ldots,d_s$. If we delete the 1's from this sequence, we obtain an invariant factor sequence for F/N; the number of zeros in this sequence is the torsion-free rank of F/N, and the initial non-zero members of the sequence form a sequence of torsion invariants for F/N (and so for A).

Now if the elements $n_i = \sum r_{ji} f_j$ are linearly independent and so constitute a basis for N, the results of Chapter 7, §3 tell us what to do. We let $R = (r_{kl})$ be the matrix of the basis $\{n_i\}$ with respect to $\{f_j\}$, find invertible matrices X and Y over $\mathbf{Z}$ such that $X^{-1}RY$ is an invariant factor matrix for R, and use X and Y to determine the new bases in F and N respectively.

In fact, the same process works even if the n_i are not linearly independent. For suppose simply that $\{n_1,\ldots,n_t\}$ generates N, let $Y = (y_{kl})$ be any invertible $t \times t$ matrix over $\mathbf{Z}$, and set $n'_i = \sum_{j=1}^{t} y_{ji} n_j$ for $i = 1,\ldots,t$. If $Y^{-1} = (\hat{y}_{kl})$, then

$$\sum \hat{y}_{ji} n'_j = \sum \hat{y}_{ji} y_{kj} n_k = \sum y_{kj} \hat{y}_{ji} n_k = \sum \delta_{ki} n_k = n_i.$$

Thus the subgroup (or **Z**-submodule) generated by the n'_i contains the n_i and so is N.

In the general case then, let $R = (r_{kl})$ be the $s \times t$ matrix of the set $\{n_i\}$ with respect to $\{f_j\}$. By 7.10 there exist invertible $s \times s$ and $t \times t$ matrices $X = (x_{kl})$ and $Y = (y_{kl})$ respectively, both over $\mathbf{Z}$, such that

$$X^{-1} R Y = \text{diag}(d_1,\ldots,d_u)$$

with $d_1|\cdots|d_u$ ($u = \min\{s,t\}$). Not only do the matrices X and Y exist, but we have a systematic method of finding them. Let

$$f'_i = \sum_{j=1}^{s} x_{ji} f_j \quad (i = 1,\ldots,s),$$

$$n'_i = \sum_{j=1}^{t} y_{ji} n_j \quad (i = 1,\ldots,t).$$

Then $\{f'_1,\ldots,f'_s\}$ is a basis of F, and by the above remark $\{n'_1,\ldots,n'_t\}$ generates N. By the argument of p. 108 the matrix of the set $\{n'_i\}$

with respect to $\{f'_j\}$ is $\mathrm{diag}(d_1,\ldots,d_u)$. We now consider two cases according as $t \leqslant s$ or $s < t$.

Case 1. $t \leqslant s$. Then $u = t$ and $n'_i = d_i f'_i$ for $i = 1,\ldots, t$. We then have

$$N = \mathbf{Z}(d_1 f'_1) \oplus \cdots \oplus \mathbf{Z}(d_t f'_t) = \mathbf{Z}(d_1 f'_1) \oplus \cdots \oplus \mathbf{Z}(d_s f'_s),$$

where we define $d_{t+1} = \cdots = d_s = 0$. A sequence of invariant factors for F/N is then obtained by striking out the initial 1's from the sequence $d_1,\ldots, d_t, 0,\ldots, 0$ (with $s - t$ zeros).

Case 2. $s < t$. Then $u = s$, and we have $n'_i = d_i f'_i$ for $i = 1,\ldots, s$; $n'_i = 0$ for $i > s$. Then $n'_{s+1},\ldots, n'_t$ can be suppressed, $N = \mathbf{Z}(d_1 f'_1) \oplus \cdots \oplus \mathbf{Z}(d_s f'_s)$, and $d_1,\ldots, d_s$ becomes a sequence of invariant factors for F/N after the 1's have been deleted.

We thus have a systematic scheme of calculation which will tell us, in a finite number of steps, the torsion-free rank and torsion invariants of any given Abelian group presented by a finite number of generators and relations. Such a systematic scheme of calculation is called an *algorithm*. The situation for Abelian groups contrasts strikingly with that for general groups – it is known that no algorithm can exist which will always determine, after finitely many steps, whether a given (non-Abelian) group, presented by a finite number of generators and relations, is the unit group.

Worked Examples. 1. Determine the torsion-free rank and torsion invariants of the Abelian group $B = \langle a,b : 2a + 3b = a - 7b = 0 \rangle$ mentioned above.

Here the 'matrix of relations' is

$$\begin{bmatrix} 2 & 1 \\ 3 & -7 \end{bmatrix}.$$

By an obvious scheme of reduction we have

$$\begin{bmatrix} 2 & 1 \\ 3 & -7 \end{bmatrix} \to \begin{bmatrix} 1 & 2 \\ -7 & 3 \end{bmatrix} \to \begin{bmatrix} 1 & 2 \\ 0 & 17 \end{bmatrix} \to \begin{bmatrix} 1 & 0 \\ 0 & 17 \end{bmatrix}.$$

Therefore $B = C_1 \oplus C_{17} = C_{17}$. The torsion-free rank is 0 and there is a single-torsion invariant, namely 17.

2. Determine the torsion-free rank and torsion invariants of $C = \langle a,b,c : a + b + c = 3a + b + 5c = 0 \rangle$.

FINITELY-GENERATED ABELIAN GROUPS

The matrix of relations is now

$$\begin{bmatrix} 1 & 3 \\ 1 & 1 \\ 1 & 5 \end{bmatrix} \to \begin{bmatrix} 1 & 3 \\ 0 & -2 \\ 0 & 2 \end{bmatrix} \to \begin{bmatrix} 1 & 0 \\ 0 & -2 \\ 0 & 2 \end{bmatrix} \to \begin{bmatrix} 1 & 0 \\ 0 & 2 \\ 0 & 0 \end{bmatrix}.$$

Here we are in the case $s > t$. An invariant factor sequence for C is therefore 2, 0; hence $C = C_2 \oplus C_0$. Thus C has torsion-free rank 1 and the single torsion invariant 2.

3. Determine a direct decomposition in the style of 10.1 for $A = \langle a,b,c : 7a + 4b + c = 8a + 5b + 2c = 9a + 6b + 3c = 0 \rangle$.

The matrix of relations here is

$$R = \begin{bmatrix} 7 & 8 & 9 \\ 4 & 5 & 6 \\ 1 & 2 & 3 \end{bmatrix}$$

which by coincidence is the one we reduced on p. 120. The invariant factors of this matrix are 1, 3, 0, and so our group is $C_3 \oplus C_0$. However, this does not tell us how to find actual *subgroups* of A giving such a decomposition. These may be found as follows.

Let F be a free Abelian group with basis $\{x,y,z\}$, let ϵ be the epimorphism sending $x \to a$, $y \to b$, $z \to c$, and let $N = \ker \epsilon$. Then N is generated by the elements $7x + 4y + z$, $8x + 5y + 2z$ and $9x + 6y + 3z$, the matrix of these elements with respect to $\{x,y,z\}$ being the matrix R above. If U and Y are invertible 3×3 matrices such that $U^{-1}RY = \text{diag}(1,3,0)$ and $\{x',y',z'\}$ is the basis of F whose matrix with respect to $\{x,y,z\}$ is U, then we know on general grounds from Chapter 7, §3, that $N = \langle x' \rangle \oplus \langle 3y' \rangle$ (where we use $\langle S \rangle$ to denote the group generated by a set S, rather than $\mathbf{Z}S$). Therefore F/N is the direct sum of a cyclic group of order 3 generated by $y' + N$ and one of infinite order generated by $z' + N$. Considering the diagram where ν is the natural homomorphism and ψ is the unique isomorphism which makes this diagram commute, we reach the conclusion that, as ψ is an isomorphism, A is the direct sum of a cyclic group of order 3 generated by $\psi(y' + N) = \psi\nu(y') = \epsilon(y')$ and an infinite cyclic group generated by $\epsilon(z')$.

We therefore need to find the matrix U. U^{-1} is the matrix X given on p. 121. However, rather than calculate X^{-1} directly, we recall that X was obtained by applying a certain sequence of row operations to 1_3. X^{-1} will therefore be obtained by applying the inverses of these operations, in the opposite order, to 1_3. The inverse of $R_i + cR_j$ (in the notation of p. 120) is $R_i - cR_j$, of $R_i \leftrightarrow R_j$ is itself, and of uR_i is $u^{-1}R_i$. Using these operations we then obtain

$$U = X^{-1} = \begin{bmatrix} 7 & 2 & 1 \\ 4 & 1 & 0 \\ 1 & 0 & 0 \end{bmatrix}.$$

Therefore $x' = 7x + 4y + z$, $y' = 2x + y$, $z' = z$ is the required basis of F, and $\epsilon(y') = 2a + b$, $\epsilon(z') = c$. Hence $A = \langle 2a + b \rangle \oplus \langle c \rangle$, these cyclic subgroups being of order 3 and of infinite order respectively. The torsion subgroup of A is $\langle 2a + b \rangle$, and, although it may be presented with a different generator (e.g. $4a + 2b$), as a subgroup it is uniquely determined in such a decomposition. In contrast, the torsion-free component is not uniquely determined; for example, we have $A = \langle 2a + b \rangle \oplus \langle 2a + b + c \rangle$, and clearly the second component is different from $\langle c \rangle$.

Exercises for Chapter 10

1. Classify the Abelian groups of order (a) 40, (b) 136, (c) 1080, (d) 1001. In other words, for each of the given orders write down a list containing exactly one representative from each isomorphism class of Abelian groups of that order. Give the torsion and primary invariants of each of your groups.

2. Find the order of the Abelian group $\langle a, b : 3a + 6b = 9a + 24b = 0 \rangle$ and give its torsion invariants.

3. Determine the torsion-free rank and torsion invariants of $\langle a, b, c : 2a + b = 3a + c = 0 \rangle$.

FINITELY-GENERATED ABELIAN GROUPS

4. Let
$$A = \langle a,b,c : -4a + 2b + 6c = -6a + 2b + 6c = 7a + 4b + 15c = 0 \rangle.$$

   Show that $|A| = 12$. Find elements u and v in A, of orders 2 and 6 respectively, such that $A = \langle u \rangle \oplus \langle v \rangle$.

5. Invent some more numerical examples for yourself if you feel it necessary.

6. Find indecomposable subgroups whose direct sum is the additive group of $\mathbf{Z}_n$ for $n = 10, 12, 30, 252$. The elements of each subgroup should be listed explicitly, using the notation $0, 1, \ldots, n-1$ (with square brackets if you prefer) to describe them, e.g. $\mathbf{Z}_6 = \{0,2,4\} \oplus \{0,3\}$. Lemma 8.11 will be useful.

7. Suppose that $R = (r_{kl})$ is an invertible $s \times s$ matrix over $\mathbf{Z}$. What can you say about the group
$$\langle a_1, \ldots, a_s : \sum_{j=1}^{s} r_{ji} a_j = 0 \text{ for } i = 1, \ldots, s \rangle ?$$

8. Let r, s, t, u be integers and let $d = ru - st$. Describe the structure of $\langle a, b : ra + tb = sa + ub = 0 \rangle$ when d is (i) 1, (ii) 2, (iii) a prime, (iv) 0.

9. Let A be an Abelian group presented with s generators and t relations, where $t < s$. Show that the torsion-free rank of A is at least $s - t$.

10. Let A be a finite Abelian group all of whose elements have order a power of a fixed prime p. Prove that $|A|$ is a power of p.

11. Let p be a prime, and let A be a non-trivial cyclic group of p-power order. Prove that the equation $px = 0$ has p solutions in A. Prove that, if A is the direct sum of s non-trivial cyclic groups of p-power order, then the equation $px = 0$ has p^s solutions in A.

12. Let **k** be a finite field and let **k*** be the multiplicative group **k**∖{0}. By interpreting the result of Exercise 11 in multiplicative notation, prove that each primary component of **k*** is cyclic. Deduce that **k*** is cyclic.

13**. Let p be a prime and let A be a finite Abelian group of order a power of p, with primary invariants $p^{\alpha_1},\ldots,p^{\alpha_s}$, where $\alpha_1 \leqslant \cdots \leqslant \alpha_s$. Let B be a subgroup of A. Show that B is a direct sum of cyclic subgroups of orders $p^{\beta_1},\ldots, p^{\beta_s}$, where $\beta_1 \leqslant \cdots \leqslant \beta_s$ and $\beta_i \leqslant \alpha_i$ for each i (some of the β_i may be zero). (*Hint*: use induction on $\sum \alpha_i$. The result of Chapter 9, Exercise 2 may be useful.) What can you say about the relationship between the torsion invariants of a subgroup of a finite Abelian group and those of the whole group?

CHAPTER ELEVEN

Linear transformations, matrices and canonical forms

Let V denote (throughout this chapter) a vector space of dimension $n > 0$ over a field $\mathbf{k}$. We know from the elementary theory of linear algebra that a given linear transformation α of V into itself can be represented by various $n \times n$ matrices over $\mathbf{k}$. In fact, there is a unique matrix corresponding to each choice of basis of V (see the discussion in §2 of Chapter 7). In a practical situation it will obviously be desirable to know how to choose a 'nice' basis which makes the matrix of α as simple as possible, or in other words, roughly as much like a diagonal matrix as possible. The problem of choosing such bases is the one with which we shall now concern ourselves. We attack it by making V into a $\mathbf{k}[x]$-module via α (as explained on p. 72), noting that $\mathbf{k}[x]$ is a PID, and applying the powerful decomposition theorems which we have developed in Chapter 8. The solution leads to a classification of the elements α in $\text{End}_\mathbf{k} V$ to within a certain equivalence relation.

1. Matrices and linear transformations

Notation. Let $\mathbf{v} = \{v_1, \ldots, v_n\}$ be a basis of V and let $\alpha \in \text{End}_\mathbf{k} V$, the ring of linear transformations of V. We write $M(\alpha, \mathbf{v})$ for the matrix (a_{kl}) of α with respect to $\mathbf{v}$, defined as in Chapter 7, §2 by

$$\alpha(v_i) = \sum_{j=1}^{n} a_{ji} v_j \quad \text{for } i = 1, \ldots, n. \tag{1}$$

If $\mathbf{v}^* = \{v_1^*, \ldots, v_m^*\}$ is a finite subset of V, then we write $M(\mathbf{v}^*, \mathbf{v})$ for the matrix (b_{kl}) of $\mathbf{v}^*$ with respect to $\mathbf{v}$, defined by

$$v_i^* = \sum_{j=1}^{n} b_{ji} v_j \quad \text{for } i = 1, \ldots, m. \tag{2}$$

Let $\alpha \in \mathrm{End}_k V$, and let $\mathbf{v}$ and $\mathbf{v}^*$ be bases of V. Then as we saw in Chapter 7, §2, the matrices $A = M(\alpha, \mathbf{v})$ and $A^* = M(\alpha, \mathbf{v}^*)$ are related by the equation

$$A^* = X^{-1}AX,$$

where $X = M(\mathbf{v}^*, \mathbf{v})$, an invertible $n \times n$ matrix. Conversely, given such a matrix X we can use (2) to construct a basis $\mathbf{v}^*$ of V whose matrix with respect to $\mathbf{v}$ is X; then $M(\alpha, \mathbf{v}^*) = X^{-1}AX$. This prompts the following definition:

11.1. Definition. Let $A, B \in \mathbf{M}_n(\mathbf{k})$. We say that A is *similar* to B if and only if there exists an invertible $n \times n$ matrix X over $\mathbf{k}$ such that $B = X^{-1}AX$.

It is easy to see that similarity is an equivalence relation on $\mathbf{M}_n(\mathbf{k})$. By what we have said, if α is a linear transformation of V whose matrix with respect to some specified basis $\mathbf{v}$ of V is A, then the various matrices by which α can be represented with respect to the various bases of V are precisely the matrices similar to A. The problem we mentioned in the introduction is therefore equivalent to the following:

Given an $n \times n$ matrix A over $\mathbf{k}$, find a matrix A^* of simple form which is similar to A, and find an invertible matrix X over $\mathbf{k}$ such that $X^{-1}AX = A^*$.

For, suppose that the original problem of finding a nice basis for an endomorphism has been solved, and that we are given some $n \times n$ matrix A over $\mathbf{k}$. Using (1) we make A act as an endomorphism α of some prototype n-dimensional vector space V (usually the space $\mathbf{k}^n$ of n-tuples with entries in $\mathbf{k}$) with respect to a basis $\mathbf{v}$ (usually the basis $\{e_1, \ldots, e_n\}$, where e_i has a 1 in the i-th place and zeros elsewhere). Our solution then tells us how to choose a basis $\mathbf{v}^*$ of V such that $M(\alpha, \mathbf{v}^*)$ has the desired simple form. But $M(\alpha, \mathbf{v}^*) = X^{-1}AX$, where $X = M(\mathbf{v}^*, \mathbf{v})$, and we have solved the second problem. Arguing in the reverse direction shows that by solving the second problem we also solve the one we began with.

The second problem is important in many areas of pure and applied mathematics. We will confine ourselves here to mentioning one situation in which it crops up in group theory. Let $G = \mathbf{GL}_n(\mathbf{k})$ denote the multiplicative group of all $n \times n$ invertible matrices over $\mathbf{k}$ – this can be thought of as the group of units of $\mathbf{M}_n(\mathbf{k})$.

TRANSFORMATIONS, MATRICES AND FORMS

Then two elements of G are similar if and only if they are conjugate elements of G. Therefore solving the second problem enables us to find in each conjugacy class of G a matrix of a particularly simple form, thereby obtaining information about the conjugacy classes of G, in fact, a classification of them. Such information is valuable in many contexts, particularly so in the theory of group representations.

2. Invariant subspaces

We have already mentioned invariant subspaces on p. 75; we recall that, if $\alpha \in \mathrm{End}_k V$, then an α-invariant subspace of V is a subspace U of V satisfying the condition $\alpha(U) \subseteq U$. These are of great relevance to our problem for the following reason:

11.2. Lemma. *Let $\alpha \in \mathrm{End}_k V$, and suppose $V = V_1 \oplus \cdots \oplus V_k$, where each V_i is α-invariant. Let $\mathbf{v}^{(i)}$ be a basis of V_i for $i = 1,\ldots,k$ and let $\mathbf{v} = \bigcup_{i=1}^{k} \mathbf{v}^{(i)}$. Then $\mathbf{v}$ is a basis of V, and $M(\alpha, \mathbf{v})$ has the form*

$$A = \begin{bmatrix} A_1 & 0 & \cdots & 0 \\ 0 & A_2 & & \vdots \\ \vdots & & \ddots & 0 \\ 0 & \cdots & 0 & A_k \end{bmatrix}$$

where the blocks A_i are diagonally placed, $A_i = M(\alpha|_{V_i}, \mathbf{v}^{(i)})$, and the entries of A outside the blocks A_i are all zero. A_i is $n_i \times n_i$, where $n_i = \dim V_i$.

Conversely, if the matrix of α with respect to some basis $\mathbf{v}$ of V has the above form, then V splits up as a direct sum of k α-invariant subspaces as above.

Proof. We assume that the reader is familiar with the direct sum of subspaces. In any case, this is just the direct sum of **k**-submodules if we think of V as a **k**-module.

Let $\mathbf{v}^{(i)} = \{v_{j_{i-1}+1},\ldots,v_{j_i}\}$, where $j_0 = 0$, $j_k = n = \dim V$, and $j_i - j_{i-1} = n_i = \dim V_i$. Then $\mathbf{v} = \{v_1,\ldots,v_n\}$. Since $V = \sum V_i$, any element $x \in V$ can be expressed in the form $x = x_1 + \cdots + x_k$ with $x_i \in V_i$. But x_i can then be expressed as a linear combination of the elements of $\mathbf{v}^{(i)}$, and so x can be expressed as a linear combination of elements of $\mathbf{v} = \bigcup_{i=1}^{k} \mathbf{v}^{(i)}$. If $\sum_{j=1}^{n} \lambda_j v_j = 0$ with $\lambda_i \in \mathbf{k}$, then we obtain $y_1 + \cdots + y_k = 0$, where $y_i = \sum \lambda_j v_j$ summed from $j = j_{i-1} + 1$ to j_i, that is over the basis elements of V_i. Since the sum $V_1 \oplus \cdots \oplus V_k$ is direct, we obtain $y_i = 0$ for $1 \leq i \leq k$. Since $\mathbf{v}^{(i)}$ is a basis of V_i, this gives $\lambda_j = 0$ for $j_{i-1} + 1 \leq j \leq j_i$ and so for all j. Therefore $\mathbf{v}$ is a basis of V.

Suppose $j_{i-1} + 1 \leq j \leq j_i$. Then $v_j \in V_i$, and so $\alpha(v_j) \in V_i$. Therefore $\alpha(v_j)$ is a linear combination of elements of $\mathbf{v}^{(i)}$, that is, $\alpha(v_j) = \sum_{l=1}^{n} a_{lj} v_l$, where $a_{lj} = 0$ unless $j_{i-1} + 1 \leq l \leq j_i$. Thus, if $A = M(\alpha, \mathbf{v})$, then in columns $j_{i-1} + 1, \ldots, j_i$ of A the only non-zero entries that can appear are in rows $j_{i-1} + 1, \ldots, j_i$, giving a block A_i as described. Clearly A_i is the matrix of the restriction $\alpha|_V$ of α to V_i with respect to $\mathbf{v}^{(i)}$, and A_i is $n_i \times n_i$.

The converse is proved by reversing the argument and will be left to the reader.

Notation. 1. Let $V_1, \ldots, V_k$ be subspaces of V such that $V = V_1 \oplus \cdots \oplus V_k$, and for each i let α_i be an element of $\mathrm{End}_k V_i$. We define an element $\alpha \in \mathrm{End}_k V$ as follows: if $v \in V$, write $v = v_1 + \cdots + v_k$ with $v_i \in V_i$, noticing that the v_i are uniquely determined by v, and define

$$\alpha(v) = \alpha_1(v_1) + \cdots + \alpha_k(v_k).$$

It is easy to verify that α is a linear transformation of V; α is called the *direct sum* of $\alpha_1, \ldots, \alpha_k$ and is written $\alpha = \alpha_1 \oplus \cdots \oplus \alpha_k$. The notation $\alpha = \alpha_1 \oplus \cdots \oplus \alpha_k$ will therefore always imply that α_i is a linear transformation of a subspace V_i of V and $V = V_1 \oplus \cdots \oplus V_k$.

2. The matrix A having the form given in 11.2 is called the *diagonal sum* of the matrices A_i and is written $A = A_1 \oplus \cdots \oplus A_k$.

Lemma 11.2 brings out the correspondence between decompositions of α as a non-trivial direct sum of linear transformations on subspaces of V and matrices for α which are a non-trivial diagonal sum of smaller matrices.

3. V as a $\mathbf{k}[x]$-module

We have already discussed at some length how, given a linear transformation α of V, we can make V into a $\mathbf{k}[x]$-module via

TRANSFORMATIONS, MATRICES AND FORMS

α (see Example 4 on p. 72). If $f = a_0 + a_1 x + \cdots + a_r x^r \in \mathbf{k}[x]$ and $v \in V$, then by definition

$$fv = f(\alpha)(v) = a_0 v + a_1 \alpha(v) + \cdots + a_r \alpha^r(v).$$

As we remarked, different choices of α correspond to different $\mathbf{k}[x]$-module structures on V, but we shall be dealing with a fixed α throughout.

Furthermore, the $\mathbf{k}[x]$-submodules of V are precisely the α-invariant subspaces of V (see Example 5 on p. 75). Therefore a decomposition of V as a direct sum of $\mathbf{k}[x]$-submodules is the same thing as a decomposition of V as a direct sum of α-invariant subspaces, and we can read off such decompositions from the results of Chapter 8 once we have noticed that V is FG as $\mathbf{k}[x]$-module. First, however, we draw attention to some special properties of $\mathbf{k}[x]$ which make it more pleasant to deal with than a general PID or even a general ED.

Remark. We can, of course, also think of V as a $\mathbf{k}$-module. However, it is convenient to continue to use 'vector space' terminology for the ordinary vector space structure of V, reserving the 'module' terminology for the $\mathbf{k}[x]$-module structure which we have just introduced.

11.3. Definition. A non-zero polynomial $f \in \mathbf{k}[x]$ is called *monic* if its highest coefficient is 1, that is if f has the form

$$f = a_0 + a_1 x + \cdots + a_{r-1} x^{r-1} + x^r \quad (a_i \in \mathbf{k}, r \geq 0).$$

11.4. Lemma. *Any non-zero polynomial in $\mathbf{k}[x]$ is associated to a unique monic polynomial. In particular, no two distinct monic polynomials are associated.*

Proof. We recall that the units of $\mathbf{k}[x]$ are the elements of $\mathbf{k}^*$, commonly referred to as the non-zero scalars or the non-zero constants. Thus two polynomials are associated if and only if one is a non-zero scalar multiple of the other. If the polynomials in question are both monic, then consideration of the terms of highest degree shows that the scalar in question must be 1, so that the polynomials are equal. Furthermore, each non-zero polynomial is associated to a monic polynomial, namely that obtained by dividing through by its highest coefficient.

The monic polynomials play a rôle similar in some ways to that played by the positive integers in the last chapter. If m is an element of a $\mathbf{k}[x]$-module M and J is the order ideal of m, then, as $\mathbf{k}[x]$ is a PID, we have $J = \mathbf{k}[x]f$ for some $f \in \mathbf{k}[x]$. By 4.4 (iv) the various elements by which J can be generated are exactly the associates of f. So if $J \neq \{0\}$, there is a *unique* monic polynomial which generates J, and we can speak of *the* order of m, meaning this unique monic polynomial. If $J = \{0\}$, no ambiguity arises about the order of m anyway. This is analogous to taking the order of an element of an Abelian group to be the unique positive generator of its order ideal.

As $\mathbf{k}[x]$ is a PID, the primes of $\mathbf{k}[x]$ are the irreducible polynomials or irreducible elements in the usual sense. It is often convenient to deal with monic irreducibles because distinct monic irreducibles are never associated. Monic irreducibles are similar to positive primes in $\mathbf{Z}$ in this respect. The unique factorization theorem in $\mathbf{k}[x]$ can now be stated in the following stronger form:

If $0 \neq f \in \mathbf{k}[x]$, then f can be expressed in the form $f = ap_1 \ldots p_r$, where a is a non-zero scalar, the p_i are monic irreducible polynomials and $r \geqslant 0$. In such a factorization the scalar a is uniquely determined and the monic irreducibles $p_1, \ldots, p_r$ are determined up to the order in which they occur.

Before going on to apply the decomposition theorems of Chapter 8 to the $\mathbf{k}[x]$-module V we must check that V is indeed FG.

11.5. Lemma. *With the notation so far introduced, V is an FG torsion module over $\mathbf{k}[x]$.*

Proof. Let $\{v_1, \ldots, v_n\}$ be a basis of V. Then any element $v \in V$ can be expressed in the form $v = \sum a_i v_i$ with $a_i \in \mathbf{k}$. Since we can view the elements of $\mathbf{k}$ as constant polynomials, v is therefore a $\mathbf{k}[x]$-linear combination of $v_1, \ldots, v_n$, and so these elements generate V as $\mathbf{k}[x]$-module.

Now, as $\dim V = n$, the $n+1$ elements $v, \alpha(v), \ldots, \alpha^n(v)$ are linearly dependent over $\mathbf{k}$. Therefore there exist $b_0, \ldots, b_n \in \mathbf{k}$, not all zero, such that

$$b_0 v + b_1 \alpha(v) + \cdots + b_n \alpha^n(v) = 0.$$

TRANSFORMATIONS, MATRICES AND FORMS 173

Therefore $fv = 0$, where f is the non-zero polynomial $b_0 + b_1 x + \cdots + b_n x^n$. Hence V is a torsion module.

We can now apply Theorem 8.2 to V and we find that as $\mathbf{k}[x]$-module V can be expressed as a direct sum $V = V_1 \oplus \cdots \oplus V_s$, where each V_i is a non-trivial cyclic submodule of order d_i and $d_1 | \cdots | d_s$. As V is a torsion module, no torsion-free cyclic submodules occur, and so each d_i can be taken to be monic. Since by assumption no V_i is $\{0\}$, no d_i is 1. Now each V_i is an α-invariant subspace, and so from 11.2 $\alpha = \alpha_1 \oplus \cdots \oplus \alpha_s$, where $\alpha_i = \alpha|_{V_i}$.

11.6. Definition. A linear transformation α of V is called *cyclic of order f* if the associated $\mathbf{k}[x]$-module V is cyclic of order f.

Leaving closer investigation of this concept aside for the moment, we can now deduce from Theorems 8.2 and 8.5

11.7. Theorem. *Let $\alpha \in \mathrm{End}_{\mathbf{k}} V$. Then α can be expressed in the form $\alpha = \alpha_1 \oplus \cdots \oplus \alpha_s$ ($s > 0$), where*

(i) *α_i is a cyclic linear transformation whose order is a non-constant monic polynomial d_i, and*
(ii) *$d_1 | \cdots | d_s$.*
The monic polynomials arising from a decomposition of α satisfying (i) *and* (ii) *are uniquely determined by α.*

The uniqueness statement holds because any such decomposition of α corresponds to a 'torsion invariant' decomposition of the $\mathbf{k}[x]$-module V.

Similarly from 8.14 we obtain

11.8. Theorem. *Let $\alpha \in \mathrm{End}_{\mathbf{k}} V$. Then α can be expressed in the form $\alpha = \alpha_1 \oplus \cdots \oplus \alpha_r$ ($r > 0$), where each α_i is a cyclic linear transformation whose order is a power $q_i^{s_i}$ ($s_i > 0$) of a monic prime polynomial q_i.*

The set of monic prime powers arising from such a decomposition of α is uniquely determined by α up to the order in which the prime powers are written down.

Remark. The subspace V_i on which α_i acts is cyclic of prime-power order as $\mathbf{k}[x]$-module and so is indecomposable by 8.16. Therefore

V_i cannot be further decomposed as a direct sum of two nontrivial α-invariant subspaces and the decomposition of α given in 11.8 above is the 'finest' obtainable.

We now ask ourselves what it means for a linear transformation to be cyclic of order f. In order to answer this question we first remind ourselves of the properties of the *minimal polynomial* of a linear transformation α.

Let $J = \{h \in \mathbf{k}[x] : h(\alpha) = 0\}$. J thus consists of all polynomials $h = a_0 + a_1 x + \cdots + a_r x^r \in \mathbf{k}[x]$ such that $a_0 v + a_1 \alpha(v) + \cdots + a_r \alpha^r(v) = 0$ for all $v \in V$. It is easy to verify that J is an ideal of $\mathbf{k}[x]$. We claim further that $J \neq \{0\}$. For under the usual operations of addition and scalar multiplication $\mathrm{End}_{\mathbf{k}} V$ is a vector space of dimension n^2 over $\mathbf{k}$. The easiest way to see this is to use the fact that $\mathrm{End}_{\mathbf{k}} V \cong \mathbf{M}_n(\mathbf{k})$ and notice that $\mathbf{M}_n(\mathbf{k})$ is a vector space over $\mathbf{k}$ with basis the n^2 matrices E_{ij}, where E_{ij} has a 1 in the (i,j) place and zeros elsewhere. Therefore the $n^2 + 1$ elements $\iota, \alpha, \ldots, \alpha^{n^2}$ are linearly dependent over $\mathbf{k}$ (where ι denotes the identity transformation), and so $a_0 \iota + a_1 \alpha + \cdots + a_{n^2} \alpha^{n^2} = 0$, the zero transformation, for suitable coefficients $a_i \in \mathbf{k}$, not all zero. Therefore $a_0 + a_1 x + \cdots + a_{n^2} x^{n^2}$ is a non-zero polynomial in J.

It then follows by the remarks after 11.4 that J has a unique monic generator, called the minimal polynomial of α and denoted by $\min \alpha$. Note that $\min \alpha$ is a polynomial in $\mathbf{k}[x]$, and that it is uniquely determined by the following properties:

(i) $g(\alpha) = 0 \Leftrightarrow \min \alpha \mid g$, and
(ii) $\min \alpha$ is monic.

It follows from (i) that any non-zero polynomial g such that $g(\alpha) = 0$ has degree at least that of $\min \alpha$. Therefore $\min \alpha$ is also the unique monic polynomial of smallest degree which annihilates α. The reader is no doubt familiar with most of this from elementary linear algebra.

11.9. Lemma. *Let $\alpha \in \mathrm{End}_{\mathbf{k}} V$. Then α is cyclic if and only if there exists $v \in V$ such that the elements $v, \alpha(v), \alpha^2(v), \ldots$ span V (as vector space). In that case v generates V as $\mathbf{k}[x]$-module, and the order of α is the minimal polynomial of α.*

Proof. The statement that the elements $v, \alpha(v), \ldots$ span V means of course that every element of V can be expressed as a linear

TRANSFORMATIONS, MATRICES AND FORMS

combination of finitely many of these elements. Assume that is the case. Then, if $u \in V$, we have $u = a_0 v + a_1 \alpha(v) + \cdots + a_r \alpha^r(v)$ for suitable $a_i \in \mathbf{k}$, some of which may be zero, and so $u = gv$, where $g = a_0 + a_1 x + \cdots + a_r x^r \in \mathbf{k}[x]$. Therefore $V = \mathbf{k}[x]v$ is generated as $\mathbf{k}[x]$-module by v. Conversely, if $V = \mathbf{k}[x]v$, the above argument reversed shows that the elements $v, \alpha(v), \ldots$ span V.

The order f of α is then by Definition 11.6 the order of the $\mathbf{k}[x]$-module generator v of V, i.e. the unique monic generator of the ideal $\mathbf{o}(v)$. But by Remark 2 on p. 93 $\mathbf{o}(v) = \{g \in \mathbf{k}[x] : gV = \{0\}\} = \{g \in \mathbf{k}[x] : g(\alpha) = 0\}$, and the unique monic generator of this ideal is $\min \alpha$ by definition of the minimal polynomial.

The following example illustrates the concepts we have met so far.

Example. Let V_1 be a 1-dimensional vector space over $\mathbf{Q}$ with basis $\{v_1\}$, and let α_1 be the unique linear transformation of V_1 which maps $v_1 \to -v_1$. V_1 is obviously a cyclic $\mathbf{Q}[x]$-module via α_1 generated by v_1, since v_1 already spans V_1. The order of V_1 is $x + 1$, since $(x+1)v_1 = \alpha_1(v_1) + v_1 = -v_1 + v_1 = 0$ and evidently no non-zero polynomial of smaller degree sends v_1 to zero.

Let V_2 be a 2-dimensional space over $\mathbf{Q}$ with basis $\{v_2, v_3\}$, and let α_2 be the linear transformation of V_2 whose matrix with respect to that basis is

$$\begin{bmatrix} 2 & 0 \\ 1 & 2 \end{bmatrix}.$$

Thus $\alpha_2(v_2) = 2v_2 + v_3$, $\alpha_2(v_3) = 2v_3$. Since v_2 and $\alpha_2(v_2)$ are obviously linearly independent over $\mathbf{Q}$, they span V_2, and so, via α_2, V_2 is a cyclic $\mathbf{Q}[x]$-module generated by v_2. We notice that $(\alpha_2 - 2\iota)(v_2) = v_3$ and $(\alpha_2 - 2\iota)(v_3) = 0$, so that $(x-2)^2 v_2 = 0$. Since $\min \alpha_2$ therefore divides $(x-2)^2$ and $\alpha - 2\iota \neq 0$, we must have $\min \alpha_2 = (x-2)^2$ which is therefore the order of v_2.

Now let $V = V_1 \oplus V_2$, which we can think of as a space with basis $\{v_1, v_2, v_3\}$, and let $\alpha = \alpha_1 \oplus \alpha_2$. With respect to the basis just given, α has matrix

$$\begin{bmatrix} -1 & 0 & 0 \\ 0 & 2 & 0 \\ 0 & 1 & 2 \end{bmatrix}.$$

Thinking of V now as a $\mathbf{Q}[x]$-module via α, V_1 is a cyclic submodule of order $x+1$ generated by v_1 and V_2 is a cyclic submodule of order $(x-2)^2$ generated by v_2. Since these orders are relatively prime, 8.13 shows that V itself is cyclic of order $(x+1)(x-2)^2$ generated by $v_1 + v_2 = v$. The reader may easily verify that v, $\alpha(v)$ and $\alpha^2(v)$ do in fact span V. Furthermore, 8.10 tells us that V is the direct sum of a cyclic submodule of order $x+1$ generated by $(x-2)^2 v$ and one of order $(x-2)^2$ generated by $(x+1)v$. Evidently $(x-2)^2 v = (x-2)^2 v_1 = (\alpha_1 - 2\iota)^2 v_1 = 9v_1$ and $(x+1)v = (x+1)v_2 = \alpha_2(v_2) + v_2 = 3v_2 + v_3$. Clearly $9v_1$ is another generator for V_1, and it is easy to see that $3v_2 + v_3$ generates V_2 – this must indeed be the case, since V_1 and V_2 are the primary components of V and are therefore unique by 8.10. A sound understanding of this simple example will be of great benefit to the reader later.

4. Matrices for cyclic linear transformations

We now show how, by judicious choice of basis, the matrix of a cyclic linear transformation can be given various simple forms.

11.10. Lemma. *Let α be a linear transformation of V, and suppose that α is cyclic of order f. Suppose further that $V \neq \{0\}$. Let $m = \partial f$ be the degree of f and let v be a generator of V as $\mathbf{k}[x]$-module via α. Then the elements $v, \alpha(v), \ldots, \alpha^{m-1}(v)$ form a basis of V. In particular, $\partial f = \dim V$.*

Proof. Here f of course is assumed to be monic. Since $V \neq \{0\}$, $f \neq 1$, and so $\partial f = m > 0$.

We first show $\{v, \alpha(v), \ldots, \alpha^{m-1}(v)\}$ is linearly independent. Let $b_0, \ldots, b_{m-1}$ be elements of $\mathbf{k}$ such that $b_0 v + b_1 \alpha(v) + \cdots + b_{m-1} \alpha^{m-1}(v) = 0$. Then $(b_0 + b_1 x + \cdots + b_{m-1} x^{m-1})v = 0$. Therefore the order f of v divides $b_0 + b_1 x + \cdots + b_{m-1} x^{m-1}$; but this polynomial has degree at most $m-1$. Since $\partial f = m > 0$, it must be the zero polynomial. Hence $b_0 = \cdots = b_{m-1} = 0$.

Now we show that the given elements span V. Let $u \in V$. Since V is generated as $\mathbf{k}[x]$-module by v, we have $u = hv$ for some $h \in \mathbf{k}[x]$. By the Euclidean division property we can write $h = fq + r$, where $\partial r < \partial f$. Then $u = hv = (fq + r)v = qfv + rv = rv$.

Now r has degree at most $m-1$, and so has the form $r_0 + r_1 x + \cdots + r_{m-1} x^{m-1}$. Hence $u = rv = r_0 v + r_1 \alpha(v) + \cdots + r_{m-1} \alpha^{m-1}(v)$, a **k**-linear combination of the elements $v, \alpha(v), \ldots, \alpha^{m-1}(v)$. This proves the lemma.

11.11. Corollary. *With the hypotheses of* 11.10 *the matrix of α with respect to* $\{v, \alpha(v), \ldots, \alpha^{m-1}(v)\}$ *is*

$$C(f) = \begin{bmatrix} 0 & 0 & 0 & \cdots & 0 & -a_0 \\ 1 & 0 & 0 & \cdots & 0 & -a_1 \\ 0 & 1 & 0 & \cdots & 0 & -a_2 \\ \vdots & \vdots & \ddots & \ddots & \vdots & \vdots \\ 0 & 0 & \cdots & 1 & 0 & -a_{m-2} \\ 0 & 0 & \cdots & 0 & 1 & -a_{m-1} \end{bmatrix},$$

where $f = a_0 + a_1 x + \cdots + a_{m-1} x^{m-1} + x^m$.

$C(f)$ thus has 1's immediately below the diagonal, the coefficients of f with the highest coefficient suppressed and signs reversed in the last column, and zeros elsewhere.

Proof. Writing $v_i = \alpha^i(v)$ for $i = 0, 1, \ldots, m-1$ we have

$$\alpha(v_0) = 0 \cdot v_0 + 1 \cdot v_1 + 0 \cdot v_2 + \cdots + 0 \cdot v_{m-1},$$
$$\alpha(v_1) = 0 \cdot v_0 + 0 \cdot v_1 + 1 \cdot v_2 + \cdots + 0 \cdot v_{m-1},$$
$$\vdots$$
$$\alpha(v_{m-2}) = 0 \cdot v_0 + \cdots + 0 \cdot v_{m-2} + 1 \cdot v_{m-1}.$$

Now $\alpha(v_{m-1}) = \alpha^m(v) = -a_0 v - a_1 \alpha(v) - \cdots - a_{m-1} \alpha^{m-1}(v)$ since $f(\alpha)(v) = 0$. Therefore

$$\alpha(v_{m-1}) = -a_0 v_0 - a_1 v_1 - \cdots - a_{m-1} v_{m-1}.$$

The result then follows from the definition of the matrix of a linear transformation with respect to a given basis – see (1) on p. 167.

11.12. Definition. The matrix $C(f)$, which is uniquely determined by f, is called the *companion matrix* of f. (Notice that $C(f)$ is only defined for non-constant monic polynomials f.)

In the case when f is a prime power another choice of basis, giving a different matrix for α, is important. We shall only discuss what happens when the prime in question has degree 1; this often happens in practice in view of the following result.

11.13. Lemma. *In $\mathbf{C}[x]$ the primes are exactly the polynomials of degree 1.*

Proof. Let p be a prime polynomial in $\mathbf{C}[x]$. Then by definition p is not a constant, and so $\partial(p) \geqslant 1$. Hence p has a root $a \in \mathbf{C}$ – this is by well-known properties of the complex field (the so-called Fundamental Theorem of Algebra). Hence by 3.10 $(x-a)|p$. Since p is prime (and so irreducible), we must have $x - a \sim p$, and so p has degree 1. The converse is clear.

Remark. Here $\mathbf{C}$ can be replaced by any algebraically closed field. A field $\mathbf{k}$ is said to be *algebraically closed* if every element of degree $\geqslant 1$ in $\mathbf{k}[x]$ has a root in $\mathbf{k}$.

11.14. Lemma. *Let α be a cyclic linear transformation of V of order $(x - \lambda)^n$, where $\lambda \in \mathbf{k}$ and $n > 0$. Let v be a generator of V as $\mathbf{k}[x]$-module via α. Then $\{v, (\alpha - \lambda\iota)(v), \ldots, (\alpha - \lambda\iota)^{n-1}(v)\}$ is a basis of V. The matrix of α with respect to this basis is*

$$J(\lambda, n) = \begin{bmatrix} \lambda & 0 & 0 & \cdots & 0 & 0 \\ 1 & \lambda & 0 & \cdots & 0 & 0 \\ 0 & 1 & \lambda & \cdots & 0 & 0 \\ \vdots & \vdots & \ddots & \ddots & \vdots & \vdots \\ 0 & 0 & 0 & \ddots & \lambda & 0 \\ 0 & 0 & 0 & & 1 & \lambda \end{bmatrix},$$

the $n \times n$ matrix with λ's on the diagonal, 1's immediately below, and zeros elsewhere.

Proof. We already know from 11.10 that $\dim V = n$; it will therefore suffice to show that $\{v, (\alpha - \lambda\iota)(v), \ldots, (\alpha - \lambda\iota)^{n-1}(v)\}$ is linearly independent over $\mathbf{k}$. Suppose not, and choose elements $b_0, b_1, \ldots, b_{n-1} \in \mathbf{k}$, not all zero, such that

$$b_0 v + b_1(\alpha - \lambda\iota)(v) + \cdots + b_{n-1}(\alpha - \lambda\iota)^{n-1}(v) = 0.$$

Choose r such that $b_r \neq 0, b_{r+1} = \cdots = b_{n-1} = 0$. Then $gv = 0$, where $g = b_0 + b_1(x - \lambda) + \cdots + b_r(x - \lambda)^r$. Furthermore, $g \neq 0$, for the coefficient of x^r in g is $b_r \neq 0$. Since v has order $(x - \lambda)^n$, we must therefore have $(x - \lambda)^n | g$. But $\partial g = r < n$, and we have a contradiction. Hence $\{v, (\alpha - \lambda\iota)(v), \ldots, (\alpha - \lambda\iota)^{n-1}(v)\}$ is linearly independent.

TRANSFORMATIONS, MATRICES AND FORMS 179

Let $v_j = (\alpha - \lambda\iota)^j(v)$ for $j = 0, 1, \ldots, n-1$. Then $\alpha(v_j) = (\alpha - \lambda\iota)(v_j) + \lambda v_j = v_{j+1} + \lambda v_j$ if $0 \leq j < n - 1$, $\alpha(v_{n-1}) = (\alpha - \lambda\iota)(v_{n-1}) + \lambda v_{n-1} = (\alpha - \lambda\iota)^n(v) + \lambda v_{n-1} = \lambda v_{n-1}$. Thus

$$\alpha(v_0) = \lambda v_0 + v_1$$
$$\alpha(v_1) = \qquad \lambda v_1 + v_2$$
$$\vdots$$
$$\alpha(v_{n-2}) = \qquad\qquad \lambda v_{n-2} + v_{n-1}$$
$$\alpha(v_{n-1}) = \qquad\qquad\qquad \lambda v_{n-1},$$

and the matrix of α with respect to $\{v_0, \ldots, v_{n-1}\}$ is as stated.

11.15. Definition. A matrix of the form $J(\lambda, n)$ is called an *elementary Jordan λ-matrix*, or sometimes just an *elementary Jordan matrix*. $J(\lambda, n)$ is the elementary Jordan matrix associated with the polynomial $(x - \lambda)^n$.

5. Canonical forms

We are now ready to provide some answers to the problems with which we began this chapter.

11.16. Theorem. *Let α be a linear transformation of V. Then V has a basis $\mathbf{v}$ such that*

$$M(\alpha, \mathbf{v}) = C(d_1) \oplus \cdots \oplus C(d_s),$$

where $C(d_i)$ is the companion matrix of a non-constant monic polynomial d_i and $d_1 | \cdots | d_s$.

The monic polynomials occurring are uniquely determined by α.

Thus $M(\alpha, \mathbf{v})$ has the blocks $C(d_1), \ldots, C(d_s)$ strung down the diagonal. The uniqueness statement, in more expanded form, says that, if $\mathbf{u}$ is a basis of V such that $M(\alpha, \mathbf{u}) = C(g_1) \oplus \cdots \oplus C(g_r)$, where $g_1, \ldots, g_r$ are non-constant monic polynomials such that $g_1 | \cdots | g_r$, then $r = s$ and $d_i = g_i$ for $i = 1, \ldots, s$. Notice that, although we claim that the matrix form is unique, we do *not* claim that there is a unique basis of V with respect to which that form is assumed.

Proof. By 11.7 we have $\alpha = \alpha_1 \oplus \cdots \oplus \alpha_s$, where α_i is a cyclic linear transformation of order d_i acting on a subspace V_i of V and $d_1 | \cdots | d_s$. Then $V = V_1 \oplus \cdots \oplus V_s$ and $\alpha_i = \alpha|_{V_i}$. By 11.11 V_i has a basis $\mathbf{v}^{(i)}$ such that $M(\alpha_i, \mathbf{v}^{(i)}) = C(d_i)$, the companion matrix of d_i.

By 11.2 the matrix of α with respect to $\mathbf{v} = \bigcup_{i=1}^{s} \mathbf{v}^{(i)}$ is the diagonal sum of the matrices $M(\alpha_i, \mathbf{v}^{(i)})$, in other words, $C(d_1) \oplus \cdots \oplus C(d_s)$.

Now if for some other basis $\mathbf{u}$ of V we have $M(\alpha, \mathbf{u}) = C(g_1) \oplus \cdots \oplus C(g_r)$, then as in 11.2 V decomposes as $U_1 \oplus \cdots \oplus U_r$ and α decomposes as $\alpha = \beta_1 \oplus \cdots \oplus \beta_r$ where $\beta_i = \alpha|_{U_i}$ and β_i has matrix $C(g_i)$ with respect to a suitable basis of U_i. By reversing the argument of 11.11 we then see that U_i is cyclic of order g_i (as $\mathbf{k}[x]$-module) generated by the first element of the basis in question. Therefore β_i is cyclic of order g_i and it follows from 11.7 (or directly from 8.5) that $r = s$ and $d_i = g_i$ for $i = 1, \ldots, s$.

As explained at the beginning of the chapter, with any theorem about choosing matrices for linear transformations there is associated an equivalent theorem about similarity of matrices. In the case of 11.16 this is

11.17. Theorem. *Any $n \times n$ matrix A over a field $\mathbf{k}$ is similar (over $\mathbf{k}$) to a unique $n \times n$ matrix $C(d_1) \oplus \cdots \oplus C(d_s)$, where $C(d_i)$ is the companion matrix of a non-constant monic polynomial d_i and $d_1 | \cdots | d_s$.*

11.18. Definition. The matrix described in 11.16 is called the *rational canonical matrix* of α. The matrix described in 11.17 is called the *rational canonical form* of A.

Remarks. 1. The term 'rational' is applied to something which depends only on the 'rational operations' of addition, multiplication, subtraction and division and so can be constructed inside any given field.
2. There is exactly one matrix of the form $C(d_1) \oplus \cdots \oplus C(d_s)$ in each similarity class of $n \times n$ matrices over $\mathbf{k}$ – this is the force of the term 'canonical form'. To decide whether two matrices are similar we simply have to compute their rational canonical forms and check whether they coincide. Hence we have a one-one correspondence between the equivalence classes of $n \times n$ matrices over $\mathbf{k}$ under similarity and the sequences of non-constant monic polynomials $d_1, \ldots, d_s$ over $\mathbf{k}$ which satisfy

(i) $d_1 | \cdots | d_s$, and
(ii) $\sum_{i=1}^{s} \partial d_i = n$.

TRANSFORMATIONS, MATRICES AND FORMS 181

3. If we extend the notion of similarity to the ring $\text{End}_k V$ by means of the isomorphism $\text{End}_k V \cong \mathbf{M}_n(\mathbf{k})$ or, equivalently, by defining that α is similar to α' if there exists an automorphism β of V such that $\alpha' = \beta^{-1}\alpha\beta$, we obtain an analogous classification of the similarity classes of $\text{End}_k V$.

From the decomposition of a module into a direct sum of primary cyclic modules we now obtain the primary rational canonical form.

11.19. Theorem. *Let α be a linear transformation of V. Then V has a basis $\mathbf{v}$ such that*
$$M(\alpha, \mathbf{v}) = C(g_1) \oplus \cdots \oplus C(g_r),$$
where each g_i is a power $q_i^{s_i}$ ($s_i > 0$) of a monic prime polynomial q_i.

The prime powers $g_1, \ldots, g_r$ occurring are uniquely determined by α up to the order in which they occur.

The corresponding statement for matrices is

11.20. Theorem. *Any $n \times n$ matrix A over $\mathbf{k}$ is similar (over $\mathbf{k}$) to an $n \times n$ matrix of the form $C(g_1) \oplus \cdots \oplus C(g_r)$, where each g_i is a power $q_i^{s_i}$ ($s_i > 0$) of a monic prime polynomial q_i. This matrix is uniquely determined up to the order of the blocks $C(g_i)$ on the diagonal.*

Proof of 11.19. This proof is conducted in exactly the same way as that of 11.16. By 11.8 we have $\alpha = \alpha_1 \oplus \cdots \oplus \alpha_r$, where each α_i is a cyclic linear transformation whose order is a non-trivial power of a monic prime polynomial, and one then proceeds as before.

By permuting the elements of the basis $\mathbf{v}$ in 11.19 we can arrange that the companion matrices corresponding to powers q^s of the same prime polynomial q are collected together on the diagonal and written in order of increasing s (and so in order of increasing size). However, in general there is no way of specifying in which order the collected blocks corresponding to different primes should occur.

11.21. Definition. The matrix of 11.19, with the diagonal blocks arranged as just described, is called a *primary rational matrix* of α. The matrix of 11.20 with the diagonal blocks similarly arranged is called a *primary rational canonical form* for A. The prime powers occurring are often called the *elementary divisors* of α (or of A). The elementary divisors of α are thus nothing more than the primary invariants of the associated $\mathbf{k}[x]$-module. It follows from the above theorems that two $n \times n$ matrices over $\mathbf{k}$ (or two linear transformations of V) are similar if and only if they have the same set of elementary divisors.

Finally, we come to the Jordan canonical form. This is not a rational form, as its existence depends on being able to solve polynomial equations, and this cannot be done in general by rational operations. However, it can always be done over an algebraically closed field such as $\mathbf{C}$.

11.22. Theorem. *Let α be a linear transformation of an n-dimensional vector space V over the complex field $\mathbf{C}$. Then V has a basis $\mathbf{v}$ such that $M(\alpha, \mathbf{v}) = J(\lambda_1, n_1) \oplus \cdots \oplus J(\lambda_r, n_r)$, each $J(\lambda_i, n_i)$ being the elementary Jordan matrix of a prime power $(x - \lambda_i)^{n_i}$ $(n_i > 0)$.*

If with respect to some basis of V the matrix of α is a diagonal sum of elementary Jordan matrices, then these matrices are the ones above in some order.

Proof. By 11.8 we have $\alpha = \alpha_1 \oplus \cdots \oplus \alpha_r$, where each α_i is a cyclic linear transformation on a subspace V_i of V and has order a power $q_i^{n_i}$ $(n_i > 0)$ of a monic prime polynomial q_i. Then $V = V_1 \oplus \cdots \oplus V_r$ and $\alpha_i = \alpha|_{V_i}$. Now since we are working over $\mathbf{C}$, Lemma 11.13 tells us that q_i is actually linear and so has the form $x - \lambda_i$ for some $\lambda_i \in C$. By 11.14 V_i has a basis $\mathbf{v}^{(i)}$ such that $M(\alpha_i, \mathbf{v}^{(i)}) = J(\lambda_i, n_i)$, and so, if $\mathbf{v} = \bigcup_{i=1}^{k} \mathbf{v}^{(i)}$, then 11.2 gives $M(\alpha, \mathbf{v}) = J(\lambda_1, n_1) \oplus \cdots \oplus J(\lambda_r, n_r)$. Notice that the decomposition of V used here is the same as that giving the primary rational matrix of α; we get the present matrix by choosing different bases in the components of V.

The uniqueness statement is proved in the usual way. If, with respect to some basis $\mathbf{u}$ of V, $M(\alpha, \mathbf{u})$ is a diagonal sum of elementary Jordan matrices, then α decomposes as a direct sum of cyclic linear transformations whose orders are the prime powers associated with these Jordan matrices. It then follows by 11.8 that the set of prime powers occurring is the same as in the original decomposition, as required.

As usual, we have an analogous result for matrices.

11.23. Theorem. *Every $n \times n$ matrix over* **C** *is similar (over* **C**) *to a diagonal sum of elementary Jordan matrices. The elementary Jordan matrices occurring in this diagonal sum are uniquely determined up to the order of their occurrence.*

In 11.22 we can arrange, by permuting the elements of the basis **v** if necessary, that the blocks $J(\lambda, l)$ corresponding to a given value of λ are collected together and written down the diagonal in order of increasing size. Thus, if $\mu_1, \ldots, \mu_k$ are the distinct values of λ occurring, then

$$M(\alpha, \mathbf{v}) = J_1 \oplus \cdots \oplus J_k,$$

where $J_i = J(\mu_i, n_{i1}) \oplus \cdots \oplus J(\mu_i, n_{i,s_i})$ and $n_{i1} \leqslant n_{i2} \leqslant \cdots \leqslant n_{i,s_i}$. Since the field **C** is not ordered, there is no natural way of specifying in which order the J_i should occur. The J_i correspond to the decomposition of the $\mathbf{k}[x]$-module V into its primary components; the further decomposition of the J_i corresponds to decomposing each primary component of V into a direct sum of primary cyclic submodules.

11.24. Definitions. A diagonal sum of elementary Jordan λ-matrices for one and the same value of λ, taken in increasing order of size, is called a *Jordan λ-matrix*. A diagonal sum of Jordan λ-matrices with distinct values of λ is called a *Jordan matrix*. By 11.22 every linear transformation α of a vector space of finite dimension $n > 0$ over **C** can be represented with respect to a suitable basis by a Jordan matrix. Such a matrix is called a *Jordan canonical matrix* of α. A Jordan matrix similar to a given $n \times n$ matrix A over **C** is called a *Jordan canonical form* (JCF for short) of A. As we have explained, it is uniquely determined by A up to the order of occurrence of the Jordan λ-blocks on the diagonal.

Remarks. 1. Although we have only developed the theory of Jordan canonical forms over **C**, the same results hold over any algebraically closed field.
2. Our results so far are not very constructive, for they give us no idea in practice how to *calculate* the canonical forms of a given matrix or linear transformation. We shall return to this problem in the next chapter and remedy the deficiency.

6. Minimal and characteristic polynomials

We have already recalled the definition of the minimal polynomial of a linear transformation in §3. The minimal polynomial of a matrix may be defined analogously; if α is a linear transformation of V, $\mathbf{v}$ is some basis of V and $M(\alpha, \mathbf{v}) = A$, then α and A have the same minimal polynomial. This is because, for any $g \in \mathbf{k}[x]$, $g(A) = M(g(\alpha), \mathbf{v})$. Another important polynomial associated with a square matrix is its characteristic polynomial.

11.25. Definition. The *characteristic polynomial* of a square matrix A over $\mathbf{k}$ is the element $\det(x1_n - A)$ of $\mathbf{k}[x]$. It is denoted by $\operatorname{ch} A$.

Let A and α be as above. Then, if $\mathbf{u}$ is some other basis of V, we have $M(\alpha, \mathbf{u}) = X^{-1} A X$ for some invertible $X \in \mathbf{M}_n(\mathbf{k})$. Now

$$\begin{aligned}\det(x1_n - X^{-1}AX) &= \det(X^{-1}(x1_n - A)X) \\ &= (\det X)^{-1} \det(x1_n - A) \det X \\ &= \det(x1_n - A) = \operatorname{ch} A.\end{aligned}$$

In other words, matrices which represent the same linear transformation α with respect to different bases have the same characteristic polynomial. We define this polynomial to be the characteristic polynomial $\operatorname{ch} \alpha$ of α.

We now investigate how the minimal and characteristic polynomials of a linear transformation fit into the framework of this chapter.

11.26. Lemma. *Let α be a linear transformation of V, and let $C(d_1) \oplus \cdots \oplus C(d_s)$ be the rational canonical matrix of α. Then*

(i) $\min \alpha = d_s$, *and*
(ii) $\operatorname{ch} \alpha = d_1 \ldots d_s$.

Proof. (i) It is most convenient here to think in module terms. We have $V = V_1 \oplus \cdots \oplus V_s$ as $\mathbf{k}[x]$-module, where V_i is cyclic of order d_i. As $d_i | d_s$, we have $d_s V_i = \{0\}$ for $1 \leqslant i \leqslant s$, and so $d_s V = \{0\}$. Therefore $d_s(\alpha) = 0$ and $g | d_s$, where $g = \min \alpha$. On the other hand, $g(\alpha) = 0$, and so $\{0\} = g(\alpha) V_s = g V_s$. Hence $d_s | g$. Therefore $d_s \sim g$, and since d_s and g are both monic, this gives $d_s = g$.

(ii) Let $A = C(d_1) \oplus \cdots \oplus C(d_s)$. Then by definition, $\operatorname{ch} \alpha = \operatorname{ch} A$. Now $x1_n - A = (x1_{n_1} - C(d_1)) \oplus \cdots \oplus (x1_{n_s} - C(d_s))$ (where

TRANSFORMATIONS, MATRICES AND FORMS 185

d_i has degree n_i), and so ch$A = \det(x1_n - A) = \det(x1_{n_1} - C(d_1))$
$\ldots \det(x1_{n_s} - C(d_s)) = \text{ch}\, C(d_1) \ldots \text{ch}\, C(d_s)$. If $d = a_0 + a_1 x + \cdots + a_{r-1} x^{r-1} + x^r$, we have

$$\det(x1_r - C(d)) = \det \begin{bmatrix} x & 0 & . & . & . & . & a_0 \\ -1 & x & 0 & . & . & . & a_1 \\ 0 & -1 & x & 0 & . & . & a_2 \\ \vdots & \vdots & \vdots & & & & \vdots \\ 0 & . & . & . & . & x & a_{r-2} \\ 0 & 0 & . & . & . & -1 & (x + a_{r-1}) \end{bmatrix}.$$

We prove by induction on r that ch$C(d) = d$. If $r = 1$, then the above determinant is equal to $x + a_0$ as stated. If $r > 1$, we expand by the top row and obtain

ch $C(a_0 + a_1 x + \cdots + a_{r-1} x^{r-1} + x^r)$
$\quad = x\,\text{ch}\, C(a_1 + a_2 x + \cdots + a_{r-1} x^{r-2} + x^{r-1}) + a_0$,

whence the result follows by induction.

We therefore obtain ch$\alpha =$ ch$A =$ ch$C(d_1) \ldots$ ch$C(d_s) = d_1 \ldots d_s$, as stated.

11.27. Lemma. *Let V be a vector space over $\mathbf{C}$, and let α be a linear transformation of V. Let $J_1 \oplus \cdots \oplus J_k$ be a Jordan canonical matrix for α, where*

$$J_i = J(\lambda_i, n_{i1}) \oplus \cdots \oplus J(\lambda_i, n_{i, s_i}),$$

$n_{i1} \leq n_{i2} \leq \cdots \leq n_{i, s_i}$ *and the λ_i are all distinct. Then*

(i) ch$\alpha = (x - \lambda_1)^{m_1} \ldots (x - \lambda_k)^{m_k}$ *where* $m_i = \sum_j n_{ij}$, *and*
(ii) min$\alpha = (x - \lambda_1)^{n_{1, s_1}} \ldots (x - \lambda_k)^{n_k, s_k}$.

Proof. (i) Let $B = J_1 \oplus \cdots \oplus J_k$. Then by the argument just given, ch$\alpha =$ ch$B = \prod_{i,j} \text{ch}\, J(\lambda_i, n_{ij})$. Now

$$\text{ch}\, J(\lambda, r) = \det \begin{bmatrix} x - \lambda & 0 & . & . & . & 0 \\ -1 & x - \lambda & 0 & . & . & 0 \\ 0 & -1 & x - \lambda & 0 & . & 0 \\ \vdots & \vdots & . & & & \vdots \\ 0 & . & . & . & x - \lambda & 0 \\ 0 & 0 & . & . & -1 & x - \lambda \end{bmatrix}$$

$= (x - \lambda)^r$, whence the result follows.

(ii) As we have said, the polynomials $(x - \lambda_i)^{n_{ij}}$ are the primary invariants of the $\mathbf{k}[x]$-module V associated with α. By 11.26 $\min \alpha$ is the highest torsion invariant of V. This is obtained by choosing for each prime polynomial the highest power to which it occurs as a primary invariant and multiplying these powers together, and gives the stated result. (An analogous process was used in the Worked Example on pp. 155–156).

Thus the total power to which $x - \lambda$ occurs in $\operatorname{ch}\alpha$ gives the size of the total Jordan λ-block in a Jordan matrix for α, and the power to which $x - \lambda$ occurs in $\min \alpha$ gives the size of the largest elementary matrix in the Jordan λ-block.

11.28. Corollary. $\min \alpha | \operatorname{ch} \alpha$ *and* $\min A | \operatorname{ch} A$ *for any linear transformation α and square matrix A.*

This result, which is immediate from 11.26, is the well-known Cayley–Hamilton Theorem, often stated in the form 'a matrix satisfies its own characteristic polynomial'.

11.29. Corollary. *For any linear transformation α, $\min \alpha$ and $\operatorname{ch} \alpha$ have the same sets of irreducible factors.*

Proof. Since $\min \alpha | \operatorname{ch} \alpha$, every irreducible factor of $\min \alpha$ is an irreducible factor of $\operatorname{ch} \alpha$. On the other hand, with the notation of 11.26 we have $\operatorname{ch} \alpha = d_1 \ldots d_s$ which divides $(d_s)^s = (\min \alpha)^s$. Hence every irreducible factor of $\operatorname{ch} \alpha$ is a factor of $(\min \alpha)^s$ and so of $\min \alpha$.

11.30. Corollary. *If α is a linear transformation represented by a Jordan matrix J, then the diagonal entries of J are precisely the roots of $\operatorname{ch} \alpha$.*

This is immediate from 11.27. The roots of $\operatorname{ch} \alpha$ are the *characteristic roots* or *eigenvalues* of α; no doubt the reader has met these before.

Worked Examples. 1. Prove that for 3×3 complex matrices the JCF can be deduced at once from a knowledge of $\min A$ and $\operatorname{ch} A$.

For 3×3 matrices a knowledge of the size of each λ-block and of the largest elementary block is sufficient to determine the JCF, and by the result of 11.27 this information can be deduced from

TRANSFORMATIONS, MATRICES AND FORMS

ch$A$ and min$A$. We can calculate ch$A$ directly, and then determine min$A$ by testing the various polynomials which (i) divide ch$A$ (11.28) and (ii) are divisible by the distinct linear factors of ch$A$ (11.29).

Let ch$A = (x - \lambda_1)(x - \lambda_2)(x - \lambda_3)$. We consider three separate possibilities and list the JCF's in each case.

Case 1. $\lambda_1, \lambda_2, \lambda_3$ all distinct.

$$\min A = (x - \lambda_1)(x - \lambda_2)(x - \lambda_3) \to \begin{bmatrix} \lambda_1 & 0 & 0 \\ 0 & \lambda_2 & 0 \\ 0 & 0 & \lambda_3 \end{bmatrix}$$

Case 2. $\lambda_1 \neq \lambda_2 = \lambda_3$. Then ch$A = (x - \lambda_1)(x - \lambda_2)^2$ and there are two possibilities:

$$\min A = (x - \lambda_1)(x - \lambda_2) \to \begin{bmatrix} \lambda_1 & 0 & 0 \\ 0 & \lambda_2 & 0 \\ 0 & 0 & \lambda_2 \end{bmatrix}$$

$$\min A = (x - \lambda_1)(x - \lambda_2)^2 \to \begin{bmatrix} \lambda_1 & 0 & 0 \\ 0 & \lambda_2 & 0 \\ 0 & 1 & \lambda_2 \end{bmatrix}$$

Case 3. $\lambda_1 = \lambda_2 = \lambda_3$. Then ch$A = (x - \lambda_1)^3$. In this case there are three possible choices for min$A$, and each gives rise to a different JCF.

$$\min A = x - \lambda_1 \to \begin{bmatrix} \lambda_1 & 0 & 0 \\ 0 & \lambda_1 & 0 \\ 0 & 0 & \lambda_1 \end{bmatrix}$$

$$\min A = (x - \lambda_1)^2 \to \begin{bmatrix} \lambda_1 & 0 & 0 \\ 0 & \lambda_1 & 0 \\ 0 & 1 & \lambda_1 \end{bmatrix}$$

$$\min A = (x - \lambda_1)^3 \to \begin{bmatrix} \lambda_1 & 0 & 0 \\ 1 & \lambda_1 & 0 \\ 0 & 1 & \lambda_1 \end{bmatrix}$$

As an illustration we calculate the JCF of the matrix

$$A = \begin{bmatrix} 0 & 1 & 0 \\ -1 & 2 & 0 \\ -1 & 0 & 2 \end{bmatrix}.$$

We have

$$\text{ch}\, A = \det \begin{bmatrix} x & -1 & 0 \\ 1 & x-2 & 0 \\ 1 & 0 & x-2 \end{bmatrix}$$

$= x(x-2)^2 + (x-2) = (x-2)(x-1)^2$. (Characteristic polynomials do not always factorize so easily!) We are therefore in Case 2. By direct calculation we check that $(A - 2 1_3)(A - 1_3) \neq 0$, and so $\min A = (x-2)(x-1)^2$. The JCF therefore

$$\begin{bmatrix} 2 & 0 & 0 \\ 0 & 1 & 0 \\ 0 & 1 & 1 \end{bmatrix}$$

2. Let A be a square matrix over $\mathbf{C}$, and suppose that $\text{ch}\, A = (x+1)^4(x+2)^3(x-2)^4$ and $\min A = (x+1)^3(x+2)(x-2)^2$. List the possibilities for the JCF of A, and with each give the rational and primary rational canonical forms. (We often speak of *the* JCF of a matrix although this is not strictly correct.)

In our solution we will use the following facts: (i) the power to which $x - \lambda$ occurs in $\text{ch}\, A$ is the size of the λ-block in the JCF of A; (ii) the power to which $x - \lambda$ occurs in $\min A$ is the size of the largest elementary λ-block in the JCF.

Thus in the JCF of the given matrix A the (-1)-block is 4×4 and contains a 3×3 elementary Jordan block. It must therefore be the diagonal sum of 1×1 and 3×3 elementary Jordan matrices, that is

$$[-1] \oplus \begin{bmatrix} -1 & 0 & 0 \\ 1 & -1 & 0 \\ 0 & 1 & -1 \end{bmatrix}.$$

Similarly, the (-2)-block is $[-2] \oplus [-2] \oplus [-2]$. The 2-block is 4×4 and contains a 2×2 elementary block. There are two possibilities:

$$\begin{bmatrix} 2 & 0 \\ 1 & 2 \end{bmatrix} \oplus \begin{bmatrix} 2 & 0 \\ 1 & 2 \end{bmatrix}$$

and

$$[2] \oplus [2] \oplus \begin{bmatrix} 2 & 0 \\ 1 & 2 \end{bmatrix}.$$

TRANSFORMATIONS, MATRICES AND FORMS

This gives two possibilities for the JCF of A, obtained by forming the diagonal sum of the (−1)-block, the (−2)-block, and one of the 2-blocks. We ignore the trivial possibility of permuting these three blocks on the diagonal.

If we choose an 11-dimensional vector space V over $\mathbf{C}$, choose a basis of V, and let α be the linear transformation of V whose matrix with respect to that basis is A, then V turns into a $\mathbf{k}[x]$-module via α in the usual way. Each elementary Jordan λ-block of size $r \times r$ in the JCF of A corresponds to a cyclic $(x - \lambda)$-primary component of order $(x - \lambda)^r$ (a vector subspace dimension r) in a decomposition of V as a direct sum of primary cyclic submodules. Therefore the primary invariants of V in the two cases are:

Case 1. $\{x+1, (x+1)^3, x+2, x+2, x+2, (x-2)^2, (x-2)^2\}$.

Case 2. $\{x+1, (x+1)^3, x+2, x+2, x+2, x-2, x-2, (x-2)^2\}$.

The torsion invariants are then obtained as in the Worked Example on p. 155; the highest torsion invariant is obtained by selecting for each prime $x - \lambda$ the highest power to which it occurs as a primary invariant and multiplying these powers together, and so on. Thus the respective sequences of torsion invariants are:

Case 1. $x+2, (x+1)(x+2)(x-2)^2, (x+1)^3(x+2)(x-2)^2$ (i.e. $x+2, x^4 - x^3 - 6x^2 + 4x + 8, x^6 + x^5 - 7x^4 - 9x^3 + 10x^2 + 20x + 8$).

Case 2. $(x+2)(x-2), (x+1)(x+2)(x-2), (x+1)^3(x+2)(x-2)^2$ (i.e. $x^2 - 4, x^3 + x^2 - 4x - 4, x^6 + x^5 - 7x^4 - 9x^3 + 10x^2 + 20x + 8$).

The primary rational canonical form is the diagonal sum of the companion matrices of the primary invariants (of suitable order), and the rational canonical form is the diagonal sum of the companion matrices of the torsion invariants, the order in this case being uniquely determined. These are therefore:

Case 1.
Primary rational form

$$[-1] \oplus \begin{bmatrix} 0 & 0 & -1 \\ 1 & 0 & -3 \\ 0 & 1 & -3 \end{bmatrix} \oplus [-2] \oplus [-2] \oplus [-2] \oplus \begin{bmatrix} 0 & -4 \\ 1 & 4 \end{bmatrix} \oplus \begin{bmatrix} 0 & -4 \\ 1 & 4 \end{bmatrix}$$

Rational form

$$[-2] \oplus \begin{bmatrix} 0 & 0 & 0 & -8 \\ 1 & 0 & 0 & -4 \\ 0 & 1 & 0 & 6 \\ 0 & 0 & 1 & 1 \end{bmatrix} \oplus \begin{bmatrix} 0 & 0 & 0 & 0 & 0 & -8 \\ 1 & 0 & 0 & 0 & 0 & -20 \\ 0 & 1 & 0 & 0 & 0 & -10 \\ 0 & 0 & 1 & 0 & 0 & 9 \\ 0 & 0 & 0 & 1 & 0 & 7 \\ 0 & 0 & 0 & 0 & 1 & -1 \end{bmatrix}.$$

Case 2 is left to the reader.

Exercises for Chapter 11

1. Find two 4×4 matrices over $\mathbf{C}$ which have the same characteristic and minimal polynomials but are not similar.

2. Let A be a matrix over $\mathbf{C}$ such that $\operatorname{ch} A = (x+1)^6(x-2)^3$ and $\min A = (x+1)^3(x-2)^2$. List the possible JCF's for A, and in each case write down the corresponding rational and primary rational canonical forms. Do the same for

$$\operatorname{ch} A = (x+1)^7(x-1)^4(x+2),$$
$$\min A = (x+1)^3(x-1)^2(x+2).$$

3. Find the various canonical forms (over $\mathbf{C}$) of

(a) $\begin{bmatrix} 0 & -1 & 2 \\ 3 & -4 & 6 \\ 2 & -2 & 3 \end{bmatrix}$, (b) $\begin{bmatrix} 0 & 0 & 1 \\ 1 & 0 & -1 \\ 0 & 1 & 1 \end{bmatrix}$,

(c) $\begin{bmatrix} 2 & 0 & 0 & 0 \\ 3 & 2 & 0 & -2 \\ 0 & 0 & 2 & 0 \\ 0 & 0 & 2 & 2 \end{bmatrix}$, (d) $\begin{bmatrix} 1 & -2 & -1 & 0 \\ 1 & 0 & -3 & 0 \\ -1 & -2 & 1 & 0 \\ 1 & 2 & 1 & 2 \end{bmatrix}$.

(Notice that while (a) and (b) can certainly be done by the methods so far developed, (c) and (d) can only be done because the matrices have been cunningly chosen.)

4. Let A be an $r \times r$ matrix over $\mathbf{k}$, and let $f = \operatorname{ch} A$. Show that f is monic and $f(0) = (-1)^r \det A$. Deduce that a companion matrix $C(g)$ is invertible if and only if $x \nmid g$.

TRANSFORMATIONS, MATRICES AND FORMS

5. Let A be a square matrix over $\mathbf{C}$. Show that A is similar to a diagonal matrix if and only if $\min A$ has no repeated roots.

6. Let α be a linear transformation of a finite-dimensional vector space V over $\mathbf{C}$, and suppose that $\alpha^s = \iota$ for some $s > 0$. Show that V has a basis $\mathbf{v}$ such that $M(\alpha, \mathbf{v})$ is diagonal (use Exercise 5).

7. List explicitly the possible rational canonical forms for 2×2 and 3×3 matrices over the field $\mathbf{Z}_2$. Show that there are 6 similarity classes in $\mathbf{M}_2(\mathbf{Z}_2)$ and 14 similarity classes in $\mathbf{M}_3(\mathbf{Z}_2)$; by observing which of these classes correspond to invertible matrices, show further that the group $\mathbf{GL}_2(\mathbf{Z}_2)$ of invertible elements in $\mathbf{M}_2(\mathbf{Z}_2)$ has 3 conjugacy classes and that $\mathbf{GL}_3(\mathbf{Z}_2)$ has 6 conjugacy classes. Carry out the same investigation for 2×2 matrices over $\mathbf{Z}_3$.

8*. Show, by considering rational canonical forms, that for $n = 2, 3, 4$ respectively, $\mathbf{M}_n(\mathbf{Z}_p)$ has $p + p^2$, $p + p^2 + p^3$, $p + 2p^2 + p^3 + p^4$ similarity classes, and that $\mathbf{GL}_n(\mathbf{Z}_p)$ has $p^2 - 1$, $p(p^2 - 1)$, $p(p^3 - 1)$ conjugacy classes. (p is a prime.)

9. Let A be a square matrix over $\mathbf{k}$. Show that A is similar over $\mathbf{k}$ to a Jordan matrix if and only if the irreducible factors of $\min A$ in $\mathbf{k}[x]$ are all linear.

10. Describe the 2×2 matrices over $\mathbf{C}$ whose similarity classes contain only one element. Generalize your answer.

11. Let $\theta: \mathrm{End}_\mathbf{k} V \to \mathbf{M}_n(\mathbf{k})$ be the ring isomorphism obtained by choosing a fixed basis of V and setting up the correspondence described in §2 of Chapter 7. Show that the definition of similarity in $\mathrm{End}_\mathbf{k} V$ given by

   $\alpha, \alpha' \in \mathrm{End}_\mathbf{k} V$ are similar if and only if the matrices $\theta(\alpha)$ and $\theta(\alpha')$ are similar

   is independent of the choice of θ. Justify the statements made in Remark 3 on p. 181.

12. Show that in $\mathbf{Z}_3[x]$, $(x-1)^3 = x^3 - 1$. Let α be a linear transformation of a finite-dimensional vector space V over the field $\mathbf{Z}_3$, and suppose $\alpha^3 = \iota$. Show that V has a basis $\mathbf{v}$ such that $M(\alpha, \mathbf{v})$ is a Jordan matrix. List the possibilities for this matrix in the case $\dim V = 3$.

    What can be said when V is over $\mathbf{Z}_p$ (p a prime in $\mathbf{Z}$) and $\alpha^p = \iota$?

13. Let α be a linear transformation of V, and think of V as a $\mathbf{k}[x]$-module via α. Suppose $\mathrm{ch}\,\alpha = p_1^{r_1} \ldots p_k^{r_k}$, where the $p_i = p_i(x)$ are distinct monic prime polynomials in $\mathbf{k}[x]$. Let
$$q_i = \prod_{j \neq i} p_j^{r_j}.$$
Show (using 8.10) that, if $\{v_1, \ldots, v_n\}$ is a basis of V, then the elements $\{q_i(\alpha)(v_j) : j = 1, \ldots, n\}$ span the p_i-primary component V_i of V. Show also that $V_i = \ker p_i(\alpha)^{r_i}$.

14. Let M be a cyclic p-torsion module over a PID R, p being a prime in R. Suppose that $M = \sum_{i=1}^s Rx_i$, where x_i has order p^{t_i} and $t_1 \leqslant t_2 \leqslant \cdots \leqslant t_s$. Suppose further that M is known to be cyclic. Show that $M = Rx_s$.

15. Let α be a linear transformation of V. Show that α is cyclic if and only if $\mathrm{ch}\,\alpha = \min \alpha$. Show that in this case the results of Exercises 13 and 14 furnish a method for determining generators for the primary components of V and hence (if $k = \mathbf{C}$) for finding a basis $\mathbf{v}$ of V such that $M(\alpha, \mathbf{v})$ is a Jordan matrix.

    Apply this method to the linear transformation of $\mathbf{C}^4$ whose matrix with respect to the 'standard basis' $\{(1, 0, 0, 0), (0, 1, 0, 0), (0, 0, 1, 0), (0, 0, 0, 1)\}$ is
$$A = \begin{bmatrix} -2 & 0 & 0 & 0 \\ 3 & 2 & 0 & -2 \\ 0 & 0 & 2 & 0 \\ 0 & 0 & 2 & 2 \end{bmatrix},$$
and hence find a 4×4 invertible matrix X over $\mathbf{C}$ such that $X^{-1}AX$ is a Jordan canonical form for A.

CHAPTER TWELVE

Computation of canonical forms

Our aim in this chapter is to give a practical method for dealing with the following two equivalent problems:

I. Given a linear transformation α of a vector space V, to find the various available canonical matrices of α and to find bases of V which give rise to these canonical matrices.

II. Given an $n \times n$ matrix A over $\mathbf{k}$, to find the various available canonical forms of A and to find invertible $n \times n$ matrices X over $\mathbf{k}$ such that $X^{-1}AX$ assumes these canonical forms.

Problem II reduces to problem I if we take an n-dimensional vector space V over $\mathbf{k}$ and let α be the linear transformation of V whose matrix with respect to some specified basis of V is A. The matrices available for α with respect to different bases of V are then the matrices similar to A, as we have frequently explained.

1. The module formulation

We begin by considering problem I for the rational canonical matrix of α. If we think of V as a $\mathbf{k}[x]$-module via α in the usual way, then the problem reduces to that of finding a 'torsion invariant' decomposition

$$V = V_1 \oplus \cdots \oplus V_s \tag{1}$$

of V, such that each V_i is a non-trivial cyclic submodule of order $d_i \in \mathbf{k}[x]$ and $d_1 | \cdots | d_s$, and finding a $\mathbf{k}[x]$-module generator for each V_i. Corollary 11.11 tells us how to construct a basis of V_i with respect to which $\alpha|_{V_i}$ has matrix the companion matrix of d_i, and then we put these bases together to get the required basis of V.

To see how to obtain the decomposition (1) we recall the way we proved the existence of such a decomposition in Chapters 7 and 8. The reader may also find it useful at this stage to look back at Worked Example 3 at the end of Chapter 10. Let $\mathbf{v} = \{v_1, \ldots, v_t\}$ be a basis of V as vector space. Then $\mathbf{v}$ certainly generates V as $\mathbf{k}[x]$-module. Let F be a free $\mathbf{k}[x]$-module with basis $\mathbf{f} = \{f_1, \ldots, f_t\}$. (Notice that we now have two kinds of bases in circulation – vector space bases and bases of free $\mathbf{k}[x]$-modules.) Then there is a unique $\mathbf{k}[x]$-module epimorphism $\epsilon : F \to V$ sending $f_i \to v_i$ for $1 \leq i \leq t$. Let $N = \ker \epsilon$, let $\mathbf{n}$ be a basis of the $\mathbf{k}[x]$-module N, and let A_x be the matrix of $\mathbf{n}$ with respect to $\mathbf{f}$. (We use the suffix x to emphasize that the entries of A_x are polynomials in $\mathbf{k}[x]$.) Let us anticipate matters slightly by asserting that N has rank t. This means that A_x is a certain $t \times t$ matrix whose entries come from $\mathbf{k}[x]$, which is a PID and even an ED. We can therefore by elementary row and column operations reduce A_x to an invariant-factor matrix $\mathrm{diag}(c_1, \ldots, c_t)$, where $c_i \in \mathbf{k}[x]$ and $c_1 | \cdots | c_t$ (cf. Chapter 7, §5). We can then find invertible $t \times t$ matrices X and Y over $\mathbf{k}[x]$ such that

$$X^{-1} A_x Y = \mathrm{diag}(c_1, \ldots, c_t).$$

Let $\mathbf{f}^* = \{f_1^*, \ldots, f_t^*\}$ be the basis of F whose matrix with respect to $\mathbf{f}$ is X. Then $\{c_1 f_1^*, \ldots, c_t f_t^*\}$ is a basis of N – it is, in fact, the basis of N whose matrix with respect to $\mathbf{n}$ is Y (cf. Chapter 7, §3). Therefore F/N is the direct sum of the cyclic submodules generated by $f_1^* + N, \ldots, f_t^* + N$, and these have orders $c_1, \ldots, c_t$ respectively. Then from the usual diagram (cf. proof of 8.2)

where ψ is an isomorphism, we find that V is the direct sum of the cyclic submodules generated by $\epsilon(f_1^*), \ldots, \epsilon(f_t^*)$, and these have orders $c_1, \ldots, c_t$. Of these modules some at the beginning may be zero; the rest furnish the required 'torsion invariant' decomposition of V.

To make this into a practical programme we have to know how

COMPUTATION OF CANONICAL FORMS

to find $f_1^*, \ldots, f_t^*$. These are determined by the matrix X which itself depends on A_x, the matrix of $\mathbf{n}$ with respect to $\mathbf{f}$. Before we can even begin therefore, we need to find a basis for N, the kernel of ϵ.

2. The kernel of ϵ

12.1. Lemma. *Using the notation of the last section, let $A = (a_{kl}) = M(\alpha, \mathbf{v})$, and set*

$$n_i = xf_i - \sum_{j=1}^{t} a_{ji} f_j \quad \text{for } i = 1, 2, \ldots, t.$$

Then $\mathbf{n} = \{n_1, \ldots, n_t\}$ *is a basis of N. In particular, N and F have the same rank.*

Proof. We notice first that a general element $f \in F$ has the form $f = \sum_{i=1}^{t} g_i(x) f_i$ with $g_i(x) \in \mathbf{k}[x]$, and that the effect of ϵ on such an element is given by

$$\epsilon\left(\sum g_i(x) f_i\right) = \sum g_i(x) v_i = \sum g_i(\alpha)(v_i).$$

Therefore $\epsilon(n_i) = \alpha(v_i) - \sum a_{ji} v_j = 0$, since $A = M(\alpha, \mathbf{v})$. Hence each n_i belongs to N.

We now show that $\mathbf{n}$ generates N. To see this, let N^* be the submodule $\sum_{i=1}^{t} \mathbf{k}[x] n_i$ generated by $\mathbf{n}$. Then

$$N^* \subseteq N. \tag{2}$$

Let W be the set of all elements of F of the form $\sum_{i=1}^{t} c_i f_i$ with $c_i \in \mathbf{k}$, and let $F^* = N^* + W$. F^* thus consists of all elements $n^* + \sum c_i f_i$ with $n^* \in N^*$ and $c_i \in \mathbf{k}$. Now F^* is clearly an additive subgroup of F and is closed under multiplication by scalars from $\mathbf{k}$. We claim that it is actually a submodule of F. For we have $xf_i = n_i + \sum a_{ji} f_j$, and so

$$x\left(n^* + \sum c_i f_i\right) = \left(xn^* + \sum c_i n_i\right) + \sum a_{ji} c_i f_j,$$

which clearly belongs to F^*. Therefore $xF^* \subseteq F^*$. It then follows by an easy induction that $x^l F^* \subseteq F^*$ for all $l \geq 1$. Therefore, if $b_0 + b_1 x + \cdots + b_k x^k \in \mathbf{k}[x]$ and $f \in F^*$, then

$$(b_0 + b_1 x + \cdots + b_k x^k) f = b_0 f + b_1(xf) + \cdots + b_k(x^k f) \in F^*,$$

which is therefore a submodule. Since F^* contains $f_1,\ldots,f_t$, we must have $F^* = F$.

Now let u be an arbitrary element of N. Then, as $u \in F^*$, we have $u = n^* + \sum c_i f_i$ for suitable $c_i \in \mathbf{k}$, $n^* \in N^*$. Therefore using (2) we get $0 = \epsilon(u) = \epsilon(n^*) + \sum c_i \epsilon(f_i) = \sum c_i v_i$. But the v_i are linearly independent elements of V, and so $c_i = 0$ for $1 \leqslant i \leqslant t$, giving $u = n^* \in N^*$. Hence $N = N^*$, as claimed.

It remains to show that $\mathbf{n}$ is linearly independent. This can be deduced from the fact that F/N is a torsion module or proved directly as follows. Suppose that $\sum h_i(x) n_i = 0$. Then substituting for the n_i we obtain

$$\begin{aligned}
0 &= \sum_i h_i(x)(xf_i - \sum_j a_{ji} f_j) \\
&= \sum_i x h_i(x) f_i - \sum_{i,j} a_{ji} h_i(x) f_j \\
&= \sum_i (x h_i(x) - \sum_j a_{ij} h_j(x)) f_i.
\end{aligned}$$

Since the f_i are linearly independent, every coefficient in this relation must vanish. Now suppose, for a contradiction, that not all the h_i are zero, and choose h_k of maximal degree, l say. Then $l \geqslant 0$, and so $xh_k(x)$ has degree $l+1$, while $\sum_j a_{kj} h_j(x)$ has degree at most l. Therefore the coefficient of f_k cannot be zero, and this is the contradiction sought.

12.2. Corollary. *The matrix A_x is*

$$\begin{bmatrix} x - a_{11} & -a_{12} & \cdots & -a_{1t} \\ -a_{21} & x - a_{22} & \cdots & -a_{2t} \\ \vdots & \vdots & & \vdots \\ -a_{t1} & -a_{t2} & \cdots & x - a_{tt} \end{bmatrix} = x 1_t - A.$$

Proof. By definition

$$n_i = -a_{1i} f_1 - a_{2i} f_2 - \cdots + (x - a_{ii}) f_i - \cdots - a_{ti} f_t.$$

12.3. Corollary. *The torsion invariants of V are the non-constant invariant factors of $x 1_t - A$.*

Proof. This follows from 12.2 and the considerations of the preceding section.

3. The rational canonical form

We now have at our disposal a method of finding the rational canonical matrix of a linear transformation (or the rational canonical form of a matrix), and to clarify matters we shall illustrate it with a numerical example. But first a remark: we notice that, in order to find a basis of V which brings an endomorphism into canonical form, we only need to know the matrix X and need not know Y (in the notation of §1). Therefore in reducing $x1_t - A$ we only need to record the row operations used; column operations need not be noted down. However, in our example we will record both kinds of operations to help the reader to follow the computations.

Worked Example. Let V be a 4-dimensional vector space over $\mathbf{Q}$ with basis $\mathbf{v} = \{v_1, v_2, v_3, v_4\}$, and let α be the linear transformation of V whose matrix with respect to $\mathbf{v}$ is

$$A = \begin{bmatrix} 2 & 0 & 0 & 0 \\ -1 & 1 & 0 & 0 \\ 0 & -1 & 0 & -1 \\ 1 & 1 & 1 & 2 \end{bmatrix}.$$

Find a basis $\mathbf{u}$ of V such that $M(\alpha, \mathbf{u})$ is the rational canonical matrix of α. Find a 4×4 invertible matrix T over $\mathbf{Q}$ such that $T^{-1}AT$ is the rational canonical form of A.

First let F be a free $\mathbf{Q}[x]$-module with basis $\mathbf{f} = \{f_1, f_2, f_3, f_4\}$ and ϵ the $\mathbf{Q}[x]$-module epimorphism sending $f_i \to v_i$ for $1 \leq i \leq 4$. Then by 12.2 $\ker \epsilon = N$ has a basis whose matrix with respect to $\mathbf{f}$ is

$$x1_4 - A = \begin{bmatrix} x-2 & 0 & 0 & 0 \\ 1 & x-1 & 0 & 0 \\ 0 & 1 & x & 1 \\ -1 & -1 & -1 & x-2 \end{bmatrix}.$$

The first step is to reduce this matrix over $\mathbf{Q}[x]$ to an invariant factor matrix. We use the notation introduced in Chapter 7, §8 for elementary row and column operations, listing at each stage of the reduction a sequence of operations which effects that stage. The reduction goes as follows.

$$\begin{bmatrix} x-2 & 0 & 0 & 0 \\ 1 & x-1 & 0 & 0 \\ 0 & 1 & x & 1 \\ -1 & -1 & -1 & x-2 \end{bmatrix} \rightarrow$$

$$\left.\begin{array}{c} R_1 \leftrightarrow R_2 \\ R_2 - (x-2)R_1 \\ R_4 + R_1 \\ C_1 - (x-1)C_2 \end{array}\right\} \begin{bmatrix} 1 & 0 & 0 & 0 \\ 0 & -(x-1)(x-2) & 0 & 0 \\ 0 & 1 & x & 1 \\ 0 & x-2 & -1 & x-2 \end{bmatrix}.$$

Now operating on the bottom right 3×3 submatrix but numbering rows and columns as rows and columns of the original matrix, we have

$$\begin{bmatrix} -(x-1)(x-2) & 0 & 0 \\ 1 & x & 1 \\ x-2 & -1 & x-2 \end{bmatrix} \rightarrow$$

$$\left.\begin{array}{c} R_2 \leftrightarrow R_3 \\ R_3 + (x-1)(x-2)R_2 \\ R_4 - (x-2)R_2 \end{array}\right\} \begin{bmatrix} 1 & x & 1 \\ 0 & x(x-1)(x-2) & (x-1)(x-2) \\ 0 & -1-x(x-2) & 0 \end{bmatrix} \rightarrow$$

$$\left.\begin{array}{c} C_3 - xC_2 \\ C_4 - C_2 \end{array}\right\} \begin{bmatrix} 1 & 0 & 0 \\ 0 & x(x-1)(x-2) & (x-1)(x-2) \\ 0 & -(x-1)^2 & 0 \end{bmatrix}.$$

We now operate on the remaining 2×2 submatrix, first bringing an element of smallest degree to the leading position.

$$\rightarrow C_3 \leftrightarrow C_4 \begin{bmatrix} (x-1)(x-2) & x(x-1)(x-2) \\ 0 & -(x-1)^2 \end{bmatrix} \rightarrow$$

$$\left.\begin{array}{c} C_4 - xC_3 \\ -1 \times C_4 \end{array}\right\} \begin{bmatrix} (x-1)(x-2) & 0 \\ 0 & (x-1)^2 \end{bmatrix}.$$

Although this is a diagonal matrix, the divisibility condition is not satisfied. We therefore proceed as follows.

$$\rightarrow \left.\begin{array}{c} R_3 + R_4 \\ C_4 - C_3 \\ C_4 \leftrightarrow C_3 \end{array}\right\} \begin{bmatrix} x-1 & (x-1)(x-2) \\ (x-1)^2 & 0 \end{bmatrix} \rightarrow$$

COMPUTATION OF CANONICAL FORMS

$$\begin{matrix} C_4 - (x-2)C_3 \\ \to \quad R_4 - (x-1)R_3 \\ -1 \times C_4 \end{matrix} \Bigg\} \begin{bmatrix} x-1 & 0 \\ 0 & (x-1)^2(x-2) \end{bmatrix}.$$

We have thus reduced $x1_4 - A$ to $\mathrm{diag}(1, 1, x-1, (x-1)^2(x-2))$, thereby finding the torsion invariants of V. By applying the sequences of row operations and column operations respectively to 1_4, we obtain 4×4 invertible matrices over $\mathbf{Q}[x]$, denoted by X^{-1} and Y respectively, such that

$$X^{-1}(x1_4 - A)Y = \mathrm{diag}(1, 1, x-1, (x-1)^2(x-2)).$$

If $\{f_1^*, f_2^*, f_3^*, f_4^*\}$ is the basis of F whose matrix with respect to $\mathbf{f}$ is X, then N has basis $\{f_1^*, f_2^*, (x-1)f_3^*, (x-1)^2(x-2)f_4^*\}$, and it follows that F/N is the direct sum of a cyclic submodule of order $x-1$ generated by $f_3^* + N$ and one of order $(x-1)^2(x-2)$ generated by $f_4^* + N$.

Therefore $V \cong F/N$ has torsion invariants $x-1, (x-1)^2(x-2)$, and so the rational canonical matrix of α is

$$C(x-1) \oplus C((x-1)^2(x-2)) = \begin{bmatrix} 1 & 0 & 0 & 0 \\ 0 & 0 & 0 & 2 \\ 0 & 1 & 0 & -5 \\ 0 & 0 & 1 & 4 \end{bmatrix} = R \quad \text{say}.$$

So far we have not needed to know the matrix X. But now to find a basis of V with respect to which α has matrix R we must indeed compute X. Recall that X^{-1} is obtained by applying the sequence of row operations used above to 1_4; X is therefore found by applying to 1_4 the inverses of these operations in the opposite order (cf. Worked Example 3, p. 163). The sequence of operations to be applied is then $R_4 + (x-1)R_3$, $R_3 - R_4$, $R_4 + (x-2)R_2$, $R_3 - (x-1)(x-2)R_2$, $R_2 \leftrightarrow R_3$, $R_4 - R_1$, $R_2 + (x-2)R_1$, $R_1 \leftrightarrow R_2$, giving

$$X = \begin{bmatrix} x-2 & -(x-1)(x-2) & -(x-2) & -1 \\ 1 & 0 & 0 & 0 \\ 0 & 1 & 0 & 0 \\ -1 & x-2 & x-1 & 1 \end{bmatrix}.$$

The co-ordinates of f_3^* and f_4^* with respect to $\mathbf{f}$ are given by the last two columns of this matrix; thus

$$f_3^* = -(x-2)f_1 + (x-1)f_4,$$
$$f_4^* = -f_1 + f_4.$$

Therefore V is the direct sum of a cyclic submodule V_1 of order $x-1$ generated by $\epsilon(f_3^*) = -(\alpha - 2\iota)(v_1) + (\alpha - \iota)(v_4) = v_2 - v_3$ and a cyclic submodule V_2 of order $(x-1)^2(x-2)$ generated by $\epsilon(f_4^*) = -v_1 + v_4$. By 11.10 V_2 has a vector space basis consisting of the elements $-v_1 + v_4$, $\alpha(-v_1 + v_4)$, $\alpha^2(-v_1 + v_4)$, which on calculation turn out to be $-v_1 + v_4$, $-2v_1 + v_2 - v_3 + v_4$, $-4v_1 + 3v_2 - 2v_3$. It then follows from 11.11 that the matrix of α with respect to the basis

$$\mathbf{u} = \{v_2 - v_3,\ -v_1 + v_4,\ -2v_1 + v_2 - v_3 + v_4,\ -4v_1 + 3v_2 - 2v_3\}$$

of V is the rational canonical matrix R.

The matrix of the basis $\mathbf{u}$ with respect to the original basis $\mathbf{v}$ is

$$T = \begin{bmatrix} 0 & -1 & -2 & -4 \\ 1 & 0 & 1 & 3 \\ -1 & 0 & -1 & -2 \\ 0 & 1 & 1 & 0 \end{bmatrix},$$

and so $T^{-1}AT = R$. This can be verified by calculation. (To avoid calculating T^{-1} check that $\det T \neq 0$ and $AT = TR$.)

4. The primary rational and Jordan canonical forms

Having obtained a basis of V with respect to which α has its rational canonical matrix we are now able without much trouble to find bases with respect to which α has a primary rational canonical matrix or a Jordan matrix. As we have pointed out, to find such bases we have to decompose V into a direct sum of primary cyclic submodules, and such a decomposition is readily obtainable (taking into account 8.11) once we have expressed V in some way as a direct sum of cyclic submodules. 11.11 and 11.14 then tell us how to choose bases in the primary cyclic summands to obtain the various canonical forms. In each case we must collect together the summands corresponding to a given prime and write them in order of increasing dimension so that the diagonal blocks will occur in the appropriate order on the diagonal.

Worked Example. With the notation of the worked example of the preceding section find bases of V with respect to which the matrix of α is (i) a primary rational canonical matrix, (ii) a Jordan

COMPUTATION OF CANONICAL FORMS

matrix. Find matrices U, W such that $U^{-1}AU$, $W^{-1}AW$ are respectively a primary rational canonical form and a JCF of A.

We have already worked out the torsion invariants of V; they are $x-1$ and $(x-1)^2(x-2)$. $V = V_1 \oplus V_2$, where V_1 is cyclic of order $x-1$ generated by $w = v_2 - v_3$ and V_2 is cyclic of order $(x-1)^2(x-2)$ generated by $u = -v_1 + v_4$. By 8.11 V_2 is the direct sum of a cyclic module V_{21} of order $(x-1)^2$ generated by $(x-2)u$ and a cyclic module V_{22} of order $x-2$ generated by $(x-1)^2u$. The primary invariants of V are thus $x-1$, $(x-1)^2$, $x-2$. Therefore a primary rational canonical matrix for α is

$$P = \begin{bmatrix} 1 & 0 & 0 & 0 \\ 0 & 0 & -1 & 0 \\ 0 & 1 & 2 & 0 \\ 0 & 0 & 0 & 2 \end{bmatrix},$$

and a JCF for α is

$$J = \begin{bmatrix} 1 & 0 & 0 & 0 \\ 0 & 1 & 0 & 0 \\ 0 & 1 & 1 & 0 \\ 0 & 0 & 0 & 2 \end{bmatrix}.$$

These can always be written down without further ado once the torsion invariants of V are known. Notice that, although a JCF is not usually available over $\mathbf{Q}$, it is available in this case because every primary invariant occurring is a power of a linear polynomial.

To obtain bases with respect to which the matrix of α assumes these forms we calculate $(x-2)u = (\alpha - 2\iota)(u)$ and $(x-1)^2 u = (\alpha - \iota)^2(u)$; these turn out to be $u_1 = v_2 - v_3 - v_4$ and $u_2 = -v_1 + v_2 - v_4$, respectively. Thus $V = V_1 \oplus V_{21} \oplus V_{22}$, and these components are cyclic of orders $x-1$, $(x-1)^2$, $x-2$ generated by w, u_1, u_2 respectively. By 11.11 a basis of V giving a primary-rational canonical matrix for α is

$$\{w, u_1, \alpha(u_1), u_2\} = \{v_2 - v_3, v_2 - v_3 - v_4, v_2 - 2v_4, -v_1 + v_2 - v_4\}.$$

The matrix of this set with respect to $\mathbf{v}$ is

$$U = \begin{bmatrix} 0 & 0 & 0 & -1 \\ 1 & 1 & 1 & 1 \\ -1 & -1 & 0 & 0 \\ 0 & -1 & -2 & -1 \end{bmatrix}.$$

and so $U^{-1}AU$ is the primary rational canonical form P; this can be checked directly.

By 11.14 a basis giving a Jordan matrix for α is $\{w, u_1, (\alpha - \iota)(u_1), u_2\}$ or $\{v_2 - v_3,\ v_2 - v_3 - v_4,\ v_3 - v_4,\ -v_1 + v_2 - v_4\}$. The matrix of this basis with respect to $\mathbf{v}$ is

$$W = \begin{bmatrix} 0 & 0 & 0 & -1 \\ 1 & 1 & 0 & 1 \\ -1 & -1 & 1 & 0 \\ 0 & -1 & -1 & -1 \end{bmatrix},$$

and one can easily check that $W^{-1}AW$ is the Jordan matrix J.

Exercises for Chapter 12

1. For the following matrices A, find invertible matrices X such that $X^{-1}AX$ assumes the various canonical forms for A. (Take the field to be $\mathbf{C}$, if necessary, to obtain the JCF.)

    (a) $\begin{bmatrix} 0 & -1 & 2 \\ 3 & -4 & 6 \\ 2 & -2 & 3 \end{bmatrix}$, (b) $\begin{bmatrix} 0 & 0 & 1 \\ 1 & 0 & -1 \\ 0 & 1 & 1 \end{bmatrix}$, (c) $\begin{bmatrix} 2 & 0 & 0 & 0 \\ 3 & 2 & 0 & -2 \\ 0 & 0 & 2 & 0 \\ 0 & 0 & 2 & 2 \end{bmatrix}$.

2. Find the JCF's of the matrices

    (a) $\begin{bmatrix} 0 & -1 & -1 & -1 \\ 1 & 2 & 1 & 1 \\ 0 & 0 & 0 & -1 \\ 0 & 0 & 1 & 2 \end{bmatrix}$, (b) $\begin{bmatrix} 0 & 1 & -2 & 1 \\ -2 & 1 & -6 & 3 \\ 2 & -3 & 0 & 1 \\ 2 & -3 & -2 & 3 \end{bmatrix}$.

3. Find the rational and primary rational canonical forms over $\mathbf{Z}_2$ of the matrix

    $$\begin{bmatrix} 1 & 1 & 1 \\ 0 & 0 & 0 \\ 1 & 1 & 0 \end{bmatrix},$$

    and prove that this matrix is not similar to a Jordan matrix over $\mathbf{Z}_2$.

COMPUTATION OF CANONICAL FORMS

4*. Let A and B be $n \times n$ matrices over a field $\mathbf{k}$. Prove that A and B are similar over $\mathbf{k}$ if and only if $x1_n - A$ and $x1_n - B$ are equivalent over $\mathbf{k}[x]$.

5*. Let V be a $\mathbf{k}[x]$-module via a linear transformation α. We give below an outline of a method, based on the proof of 9.2, which can be used to decompose V as a direct sum of primary cyclic submodules and hence to obtain the canonical forms for α. Fill in the details of each step and justify the overall argument.

(i) Find the primary components of V by the method of Chapter 11, Exercise 13. This reduces our problem to the case when V is primary.

(ii) Now assume that V is a p-torsion module, where $p = p(x)$ is a prime in $\mathbf{k}[x]$. Let $\{v_1, \ldots, v_l\}$ be any set of $\mathbf{k}[x]$-module generators of V (for example, a basis of V). Find the order p^{n_i} of each v_i. Renumber them so that $n_1 \geqslant n_i$ for all i.

(iii) Let V_1 be the submodule generated by v_1. If p^{n_1} has degree e_1, then $\mathbf{v}_1 = \{v_1, \alpha(v_1), \ldots, \alpha^{e_1 - 1}(v_1)\}$ is a basis of V_1. Delete from the generating set any v_i, $i \geqslant 2$, contained in V_1.

(iv) For each $i > 1$, find the smallest integer $m_i > 0$ such that $p^{m_i} v_i \in V_1$. By expressing $p^{m_i} v_i$ as a linear combination of the elements of $\mathbf{v}_1$, obtain an expression $p(x)^{m_i} v_i = q_i(x) v_1$. Show that $p^{m_i} | q_i$, and that, if $q_i = p^{m_i} r_i$ and $v'_i = v_i - r_i v_1$, then v'_i has order p^{m_i} and $V_1 + \mathbf{k}[x] v_i = V_1 \oplus \mathbf{k}[x] v'_i$. Notice that $m_i \leqslant n_i$.

(v) We can now assume that for each $i > 1$ $V_1 + \mathbf{k}[x] v_i = V_1 \oplus \mathbf{k}[x] v_i$. Now renumber $v_2, \ldots, v_l$ so that v_2 has order p^{n_2} with $n_2 \geqslant n_i$ for $i \geqslant 2$. Let $V_2 = \mathbf{k}[x] v_2$.

Proceeding stepwise in this manner, extend the sum $V_1 \oplus V_2$ to a direct decomposition of V into cyclic summands.

6. Use the method just developed to find a matrix X such that $X^{-1}AX$ is in JCF, where

$$A = \begin{bmatrix} 0 & 0 & -1 \\ 1 & 1 & 1 \\ 1 & 0 & 2 \end{bmatrix}$$

Try out this method on the matrices of earlier exercises.

7. Let α be an endomorphism of a vector space V (of suitable dimension over $\mathbf{C}$) such that $M(\alpha, \mathbf{v})$ is one of the matrices of Exercise 1, where $\mathbf{v}$ is some basis of V. Describe all the vectors $v \neq 0$ such that $\alpha v = \lambda v$ for some $\lambda \in \mathbf{C}$. Such vectors are called *eigenvectors* of α. (*Hint*: Lemma 8.17 may help.)

References

1. COHN, P. M. (1965). *Universal Algebra*, Harper and Row, New York.

2. HALMOS, P. (1960). *Naive Set Theory*, D. Van Nostrand, Princeton, N.J.

3. JACOBSON, N. (1951). *Lectures in Abstract Algebra*, Vol. I, D. Van Nostrand, New York.

4. KELLEY, J. L. (1955). *General Topology*, D. Van Nostrand, New York.

5. SAMUEL, P. (1968). *Unique Factorization*, American Mathematical Monthly, 75 pp. 945–952.

6. MACLANE, S. and BIRKHOFF, G. (1967). *Algebra*, Macmillan, New York.

7. ZARISKI, O. and SAMUEL, P. (1958). *Commutative Algebra*, D. Van Nostrand, Princeton, N.J.

Index

Abel, N. H., 4
Abelian group, 4, 21, Ch. 10
abusing notation, 8
addition, 3
additive (sub)group, 16
algebra over a field, 45
algebraically closed field, 178, 182, 183
Algebraic Geometry, 50
Algebraic Number Theory, 50
algorithm, 162
ascending chain condition, 68
associates, 52
associative law, 4, 11
atomic, 135
Axiom of Choice, 62
automorphism, 19, 104

bad element, 61
basis of free module, 89, 92

cancellation law, 13, 49
Cartesian product, 3
Cayley-Hamilton theorem, 186
change of basis, 105
characteristic polynomial, 184
characteristic roots, 186
classification
   of Abelian groups, 152
   of modules, 129
column operations, 111
commutative, 3

commutative diagram, 23
companion matrix, 177
component, 81
componentwise operations, 33, 35
composition of maps, 10
computing invariants, 160
congruence class modulo n, 6
conjugate quaternion, 9
constant polynomial, 39
convention for summation, 108
conversion table, 152
coordinate projections, 33
cyclic group, 152
cyclic linear transformation, 173, 176 ff
cyclic (sub)module, 77, 152

decomposition theorem, 99, Chs. 8 & 9
degree of a polynomial, 39
determinant, 45, 105
diagonal matrix, 30, 44, 109
diagonal sum of matrices, 170
diagram commutes, 23
dimension, 77, 102
direct sum
   of linear transformations, 170
   of modules, 80 ff
   of rings, 33 ff
disjoint union, 7
divides, 52

divisor, 52
divisor of zero, 13, 52

ED, 60
eigenvalues, 186
eigenvectors, 204
elementary divisors, 182
elementary Jordan λ-matrix, 179
elementary row
   and column operations, 110 ff
endomorphism
   of Abelian group, 10, 74
   of module, 78
   of ring, 19, 78
   of vector space, 72, 78
endomorphism ring, 10
epimorphism, 19
equivalence relation, 21, 53, 109, 168
equivalent matrices, 109
Euclidean algorithm, 63
Euclidean division property, 28
Euclidean domain, 60, 62 ff, 111
Euclidean function, 60, 111
external direct sum, 34, 81

factor, 52
factorization properties of Z, 51
FG, 77
field, 13
field of fractions, 51
finitely-generated
   module, 77, 85 ff
   Abelian group, Ch. 10
free Abelian group, 152
free generators, 89
free module, 89 ff
free vector space, 89, 92
fundamental theorem of algebra, 178

Gauss, theorem of, 67
Gaussian domain, 55
Gaussian integers, 7, 50, 65, 67
generates freely, 89

generators
   of Abelian group, 157
   of ideal, 27
   of (sub)module, 75, 85, 89
   of (sub)ring, 27
generators and relations, 157 ff
good element, 61
greatest common divisor, 64
group, 4, 159
group homomorphism, 18, 21, 78
group representations, 169
group theory, 168

hcf, 64
height of generating set, 140
highest common factor, 64
homomorphism,
   group, 18, 21, 78
   module, 77
   natural, 22
   ring, 18

ideal, 20, 74
identity element, 4
ideology, 69
image, 19
i-minor, 115
indecomposable module, 135
indeterminate, 39
infinite order, 88, 152
integers, 4, 28 ff
integral domain, 12, 49 ff
internal direct sum, 34, 81
invariant factor matrix, 118
invariant factors
   of matrix, 115 ff
   of module, 128
invariant subspace, 75, 169 ff
inverse, 4
inverse image, 25
invertible matrix, 104
irreducible, 55
isomorphism, 19
isomorphism theorems
   for rings, 22 ff
   for modules, 79 ff

INDEX

JCF, 183
Jordan canonical form, 183
Jordan canonical matrix, 183
Jordan λ-matrix, 183
Jordan matrix, 183

kernel, 19, 79
Kronecker delta, 45, 108

left ideal, 74, 87
left R-module, 70
length of element, 114
linearly dependent set, 90
linearly independent set, 90
linear transformation, 72

main theorem, 124
matrix of relations, 162
matrix ring, 8, 44 ff
minimal polynomial
   of linear transformation, 174
   of matrix, 184
minor, 116
module,
   cyclic, 77
   definition, 70
   examples, 71 ff
   free, 89
   finitely-generated, 77
   torsion, 87
   torsion-free, 87
   via α, 73
module homomorphism, 77
monic polynomial, 171
monomorphism, 19
morphism, 15
multiplication, 3
multiplicative function, 56
multiplicative identity, 12, 51

natural homomorphism, 21, 79
neutral element, 4
Noether, Emmy, 68
Noetherian ring, 68
non-singular matrix, 104
norm function, 56

order,
   of cyclic linear transformation, 173
   of group element, 87–88, 152
   of cyclic module, 124
ordered basis, 102
order ideal,
   of element, 87
   of cyclic module, 93
over same ring, 77

parallelogram law, 24
partial ordering, 54
periodic element, 88
PID, 59
pointwise operations, 8
polynomial function, 42
polynomial ring, 37 ff, 50
post-operator, 111
power set, 7
pre-operator, 111
presentation, 158
primary component, 133
primary cyclic module, 133
primary decomposition, 132 ff
primary invariants, 154
primary module, 133
primary rational matrix, 181
prime, 57
principal ideal, 59
principal ideal domain, 59
product of sets, 17
projections, 33, 81
p-torsion module, 133

quaternions, 9
quotient module, 79
quotient ring, 22

rank of module, 100-102
rational canonical form, 180, 197 ff
rational canonical matrix, 180, 197 ff
reduction of matrix, 111 ff
redundancy of hypotheses, 126
relations, 159

remainder theorem, 41
represent zero, 157
residue class modulo n, 6
residue class ring, 22
right R-module, 70
ring,
    additive group of, 16
    commutative, 12
    constructions, Ch. 3
    definition, 4
    direct sum, 34 ff
    examples, 5 ff
    Noetherian, 68
    non-examples, 10
    of linear transformations, 8, 103, 167
    of matrices, 8, 44
    of polynomial functions, 43
    of polynomials, 37 ff
    quotient, 22
    special classes, 12
root, 41
row operation, 110 ff
rule of thumb, 111

scalars, 171
secondary operation, 114
semigroup, 4
sequence of
    invariant factors, 118
    torsion invariants, 128, 154
similar matrices, 168
shorthand notation, 120, 156
spanning set, 92

splitting property, 106
square bracket notation, 29, 54
subgroup, 74, 152
submatrix, 116
submodule, 74 ff
subring, 15 ff

torsion
    element, 87
    module, 87 ff
torsion-free
    element, 87
    module, 87
torsion-free rank, 128, 154
torsion invariants, 128, 154
triangular matrices, 30, 44

UFD, 55
unary operation, 3
uniqueness
    of factorization, 52
    of decomposition, 127 ff
unique factorization domain, 54 ff
unit, 52, 104, 171
Universal Algebra, 80
universal property
    for direct sums, 46
    for polynomial rings, 47, 73
unordered basis, 102

vanish identically, 43

zero, 4
zero divisor, 12, 49, 52